Bayesian Inference - Recent Trends

Edited by İhsan Ömür Bucak

Published in London, United Kingdom

Bayesian Inference - Recent Trends
http://dx.doi.org/10.5772/intechopen.1000345
Edited by İhsan Ömür Bucak

Contributors
İhsan Ömür Bucak, Alexei Gilchrist, Lachlan J. Rogers, Fesih Keskin, Son T. Nguyen, Tu M. Pham, Anh Hoang, Linh V. Trieu, Trung T. Cao, Merve Veziroğlu, Erkan Veziroğlu

First published in London, United Kingdom, 2024 by IntechOpen
IntechOpen is the global imprint of INTECHOPEN LIMITED, registered in England and Wales, registration number: 11086078, 5 Princes Gate Court, London, SW7 2QJ, United Kingdom

British Library Cataloguing-in-Publication Data
A catalogue record for this book is available from the British Library

Additional hard and PDF copies can be obtained from orders@intechopen.com

Bayesian Inference - Recent Trends
Edited by İhsan Ömür Bucak
p. cm.
Print ISBN 978-1-83769-356-6
Online ISBN 978-1-83769-355-9
eBook (PDF) ISBN 978-1-83769-357-3

For EU product safety concerns:
IN TECH d.o.o., Prolaz Marije Krucifikse Kozulić 3, 51000 Rijeka, Croatia,
info@intechopen.com or visit our website at intechopen.com.

Meet the editor

İhsan Ömür Bucak graduated from İstanbul Technical University with a BSc in Electronics and Telecommunications Engineering and an MSc in Controls and Computer Engineering in 1985 and 1992, respectively. He received his Ph.D. in Electrical and Systems Engineering from Oakland University, Michigan, USA, in 2000. In addition to lecturing at various universities as an academician in various fields ranging from electrical and electronics engineering to information theory, and from control and systems engineering to computer science, he has conducted scientific research in a wide range of fields including control theory, artificial intelligence, and electric and hybrid electric vehicles. He has worked as a researcher and a professional engineer in the automotive industry for many years, gaining practical work experience and vision. He is currently active in academia as a full-time professor and department head. Dr. Bucak's main areas of expertise are nonlinear learning control and reinforcement learning, machine learning, pattern recognition and object detection, modern control theory, and automotive control with vehicle dynamics. He has published numerous scientific articles, papers, and book chapters.

Contents

Preface

Bayesian inference is a statistical inference method in which Bayes' theorem is employed to update the likelihood of a hypothesis as more evidence or information becomes available, unlike frequentist inference where the parameter is always assumed to be a fixed constant.

The main advantages of Bayesian inference over frequentist methods include the incorporation of prior knowledge into statistical analysis, the inherently probabilistic Bayesian analysis results that can provide a more intuitive understanding of the outputs, the flexible structures of Bayesian methods in handling complex models, and finally its ability to be constantly updated, which makes it ideal for sequential decision-making processes, as new data becomes available. However, it comes with some difficulties. Examples include the fact that the choice of prior has an impact on the results in case of small data, the fact that large datasets are often encountered that are computationally intensive, and the difficulty of determining a complete probabilistic model, including the priors and likelihood, and that this model requires careful assessment. Despite everything, the fact that Bayesian inference is used in a wide variety of fields can be mentioned as its greatest asset.

Ultimately, although it has its own difficulties, its ability to incorporate priors and provide probabilistic outputs gives it a tremendous edge in many fields and makes it a valuable tool, particularly in the decision-making process.

Bayesian Inference – Recent Trends is an authoritative resource that presents cutting-edge developments in Bayesian inference, a cornerstone of statistical analysis and machine learning. The book is structured as a series of comprehensive chapters, each offering a deep dive into various aspects and applications of Bayesian methods. Aimed at researchers, practitioners, and students, this book not only clarifies complex theoretical underpinnings but also demonstrates practical implementations in diverse fields.

Chapter 1, "Introductory Chapter: A Strong Come-Back of Bayesian Inference", provides a detailed introduction to the resurgence of Bayesian methods in modern analytics. It explores the fundamental concepts and advantages of Bayesian inference, particularly its proficiency in handling uncertainties and incorporating prior knowledge into model building. This chapter serves as an essential read for those seeking to understand how Bayesian inference offers a more intuitive approach to parameter estimation and uncertainty quantification compared to traditional statistical methods.

In Chapter 2, "Indirect Observation of State and Transition Probabilities", the focus shifts to the application of Bayesian techniques in hidden Markov models. The chapter presents an insightful case study on how Bayesian networks can be leveraged to infer state and transition characteristics in systems where direct observation is not feasible. It is a compelling demonstration of Bayesian inference in action, offering readers a glimpse into the method's ability to uncover hidden patterns in complex, noisy data.

Chapter 3, "Nested Sampling: A Case Study in Parameter Estimation", presents the Nested Sampling (NS) algorithm and its application in efficiently estimating parameters in high-dimensional spaces. The chapter showcases the use of NS in Direction of Arrival (DoA) estimation, emphasizing its advantages in handling multimodal distributions and complex models. This case study highlights NS's potential as a powerful alternative to conventional estimation techniques, especially in signal processing and related fields.

Chapter 4, "Bayesian Inference for Regularization and Model Complexity Control of Artificial Neural Networks in Classification Problems", focuses on Bayesian neural networks (BNNs) for classification problems with considerations of model complexity of BNNs conveniently handled by a method known as the evidence framework. In this chapter, based on the exploration of the Bayesian inference framework for MLP neural network training, it is shown that the regularization parameters (hyperparameters) and the appropriate number of hidden nodes in the network can be conveniently obtained. Therefore, a case study is presented to show that a BNN configuration based on a few nodes in the hidden layer is suitable for the incipient fault detection in power transformers. The number of hidden units mainly depends on the diagnosis criterion under consideration. This chapter also performs a comparison between suggested gas ratio limit-based methods and BNN based methods for power transformer fault diagnoses.

Finally, Chapter 5, "Performance Comparison between Naive Bayes and Machine Learning Algorithms for News Classification", introduces a Python-based news classification system, focusing on naïve Bayes algorithms for sorting news headlines into predefined categories. Four naïve Bayes variants such as Gaussian, Multinomial, Complement, and Bernoulli are evaluated and compared with classifiers such as Logistic Regression, Random Forest, Linear Support Vector Machines, Multilayer Perceptron (MLP), Decision Trees, and K-Nearest Neighbors. Ultimately, the MLP classifier achieved the greatest accuracy, underscoring its effectiveness, while Multinomial and Complement naïve Bayes has proven its robustness in news classification. Effective preprocessing of the dataset comprising BBC news headlines spanning technology, business, sports, entertainment, and politics has played a pivotal role in accurate categorization. In conclusion, this chapter gives insights into naïve Bayes algorithm performance in news classification, benefiting Natural Language Processing (NLP) and news categorization systems.

Intriguing together, these chapters offer a comprehensive overview of the contemporary landscape of Bayesian inference, illustrating its versatility and potency across different domains. *Bayesian Inference – Recent Trends* is an indispensable guide for anyone interested in the evolving field of Bayesian analysis and its myriad applications in the data-rich world of today.

İhsan Ömür Bucak
Engineering Faculty,
Department of Computer Engineering,
Şehit Bülent Yurtseven Campus,
Iğdır University,
Iğdır, Türkiye

Chapter 1

Introductory Chapter: A Strong Come-Back of Bayesian Inference

İhsan Ömür Bucak

1. Introduction

The Oxford Reference dictionary defines Bayesian inference as a statistical inference technique that estimates the probability of an event taking place in terms of the frequency of occurrence of the event, and also notes that it is based on Bayes' theorem [1].

Bayesian methods allow uncertainty to be evaluated on the basis of performed predictions. They can also show superior performance in utilizing prior knowledge and bringing robustness into the model, in principle, due to their inherent effective containment strategy in handling uncertainties related to parameter estimates [2].

Bayesian methods as a generative approach can learn not only related priors, but also adopted priors especially for hyper-parameters through training data without requiring expensive cross-validation techniques, by allowing uncertainty assessment about the (performed) predictions and the parameter estimates [2].

Answers to a state-of-the-art question of how fast, accurate, reliable and scalable inference algorithms can be strengthened in terms of design with current optimization and stochastic approximation techniques have been the subject of works conducted in recent years [2].

Bayes theorem addresses the more difficult inverse problem (posterior) that is related to any machine learning process, as opposed to the easy one which is forward problem (likelihood). In other words, the objective is to estimate or infer a model given the observed data. This is exactly the task of machine learning. Its update is based on the assumption of a prior (knowledge) and the availability of the estimated distribution of the data represented by the denominator in the Bayes theorem which is not taken into consideration in the discriminative models at all [2]. However, in general, overfitting remains one of the main challenges between machine learning and the inverse problems [2].

A prior can be placed on both parametric models consisting of a fixed number of unknown parameters and non-parametric models consisting of functions and/or a non-constant number of unknown parameters, the number of which varies only with the size of the data set. Practically, a specific prior is determined by the balancing choice between the prior itself which takes inspiration from its own strength, and the posed difficulty in the derivation of inference algorithms [2].

An important feature of Bayesian learning is its posterior distribution, which can be used to generate a point prediction and provide the amount of uncertainty of the point prediction.

The number of parameters used in non-parametric models and learned directly from the data during training is often initially very large, and then the

 IntechOpen

algorithm takes on the task of reducing them to a limited set of parameters. Here, the priors referring to an infinite number of parameters take place in the scene, and the "true" number of parameters is tried to be recovered by the respective posteriors [2].

2. The place and importance of Bayesian techniques in machine learning

Machine learning is just one part of the field of artificial intelligence, and today there are many developments using machine learning and its models. Bayesian and statistical algorithms are a subsection of machine learning.

Since measurements of dependent and independent variables are inherently noisy and imprecise, and the relationship between them is invariably non-deterministic, it proves to us that a consistent and principled way that will enable us to make meaningful reasoning in the presence of this uncertainty is only possible through probability theory and especially Bayes' rule, which explains the logic of uncertainty [3]. One of the main tasks of machine learning is to approximate the conditional probability $P(A \mid B)$ with a suitably specified model based on given sets of examples corresponding to these measurements, if we call the dependent and independent variable measurements A and B respectively.

Bayesian inference is more prominent in the modeling procedure. We have a type of parameterized model defined as follows, where, $\boldsymbol{w}$ represents a vector of all "tunable" parameters in the model as a conditional probability:

$$P(B|A) = f(A; \boldsymbol{w}). \tag{1}$$

The task of predicting the unknown B based on A is done by evaluating $f(A; \boldsymbol{w})$ such that the $\boldsymbol{w}$ parameters are optimum. Bayesian inference puts parameters, such as $\boldsymbol{w}$, in the same category of random variables as A and B. However, if the f model is made too complex, we may run the risk of overtraining the observed D data and, as a result, realizing a poor $P(B \mid A)$ distribution model [3]. Here, a given set D consists of N examples of variables A and B. Overtraining or overspecializing occurs in training algorithms, where a limited number of training examples are available. Since the training data set, which contains a large number of training samples, will perfectly represent the test set, no difference should be expected between the performance of the training set and the test set. On the other hand, since the data sets are limited in practice, it is normal for the test set performance to be worse than the training set performance. This is because the model is very specialized and cannot generalize well [4].

Machine learning methods have been extensively used for over the decades to extract useful information out of data and then use it to perform predictions. Such methods, at most, count on employing a parametric model to describe the data given and then estimates to describe the unknown model parameters are driven through an inference on an estimation technique.

Machine learning from the Bayesian point of view is receiving a lot of attention in recent years, due to its relative advantages and the additional knowledge provided by posteriors/posterior distributions.

3. The place of Bayesian techniques in deep neural networks (DNNs)

We have witnessed the return of Bayesian methods as a new source of inspiration in the design of deep neural networks, with strong ties to Bayesian models and unsupervised learning [2].

Bayesian DNN's inference steps differ from their deterministic peers in two ways: The first is that when unknown synaptic parameters or weights are defined in the form of parameterized distributions, the cost function to be optimized is expressed not by synaptic weights but by hyper-parameters describing the relevant distributions. The second is that the evidence function to be maximized is not traceable.

After 2010, the most important design problem in DNNs that dominate the machine learning field with their tremendous representation capability and their extraordinary predictive achievement for many learning tasks is pruning, which leads to the removal of unnecessary nodes and links and thus plays a major role in reducing the number of associated parameters as well as the DNN size.

4. Applications of Bayesian learning models

A wide variety of applications of Bayesian learning models attract attention today. It is certain that Bayesian learning models are particularly widespread around large and complex wireless communication systems, such as 6G, automated systems that constantly face fast-charging environments and critical decision-making processes.

In this context, we will try to briefly emphasize the diversity of applications in this field by giving some examples of innovative studies that we have selected in recent years from the present to the past.

The first of these studies was chosen from communication theory. Bayesian inference has been conducted for simulating radio channels on a newly proposed stochastic multipath model to approximate the analytically intractable posterior distribution as likelihood function by introducing a novel Markov Chain Monte Carlo technique to improve the efficiency of Monte Carlo computations [5].

Bayesian inference can also be applied in machine learning techniques for system identification. For example, Bayesian inference has found application in performing system identification by modeling posterior distributions of the model parameters. To this end, Bayesian inference has been used to improve the performance of black-box high-precision modeling on a fine-tuning training of neural networks for unknown non-linear systems using informative seismic data to evaluate the methods. Likewise, the constitution of the machine learning structures can also be used as a prediction tool in time series as is its simplest form of Bayesian inference [6].

In another similar application, Bayesian inference has been used as based on data-driven training method to train on-line the historical seismic data of Central Italy between the years 2014 and 2016 as (having) a prior knowledge for predicting three important seismological parameters in future events as likelihood, such as time, magnitude and location, and updating the posterior distribution with more recent data in order to improve forecasting under the existence of such uncertainty [7].

Bayesian statistics supports detailed and rigorous data analysis for experimental software engineering, whereas frequentist statistics has many fundamental issues. In the case of principled data analysis processes, their results are interpreted more naturally and directly related to practically important measures. This points to the closing of the gap between statistical and practical significance, which makes the real difference in practice. Bayesian statistics can help experimental software engineering establish solid foundations and achieve solid results. Some Bayesian analysis techniques can even be used to simulate scenarios that are realistic and different from those measured; this capability accelerates decision-making in real-world software engineering scenarios [8].

Bayesian inference has been used in the semiconductor industry to propose a new version of cell-aware diagnosis flow to precisely identify defect candidates of customer returns, and to confine increasing number of defects in parallel with unending effort of continued pushes for higher density within the cell structures of digital ICs. Cell-Aware test is the only way today to achieve the required low defective-parts-per-million rates for critical applications in an attempt to ensure no similar defect occurrences in the future [9].

Bayesian inference has been used as part of statistical modeling for predicting mean value of theoretical housing prices out of the given dataset, and to validate it in an attempt to apply the model further to foresee more theoretical asset price of similar large items that help economists calculate the difference between the theoretical asset price and the asset price subjected to an abrupt change under the impact of COVID-19 [10].

Rapid development of machine learning has boosted joint effort of computer scientists on workload characterization, particularly on acceleration-related workloads in deep learning, which is a leading algorithm to advance learning patterns from large mass of labeled data, or supervised learning, and on performance bottlenecks that lend themselves to optimization. In deep learning, models are trained using labeled data, while uncertainty or noise is not explicitly modeled. When we do not have enough labeled data or are trying to establish an uncertain relationship between different types of observed data, we can build a Bayesian model and use Bayesian inference to find out what we want from the data. Bayesian inference is a particularly important machine learning technique that complements deep learning in many fields. Bayesian methods often out rate deep learning in supervised learning. In particular, Bayesian inference and modeling are more successful in producing more interpretable models, especially in the case of unlabeled or limited data, using informative priors to its full extent in favor of its own influence. One of the suggested techniques in this context claims to improve Bayesian inference performance by an average of 5.8x times (by 5.8x on average) compared to naive assignment and execution of the workloads. The authors even claim that the proposed techniques can facilitate the deployment of Bayesian inference as a genetic web service [11].

Policy evaluation in on-line reinforcement learning has been considered using Gaussian processes to represent the action value function in terms of probabilistic perspective. What is meant by probabilistic perspective is the sequential update of the action value function with respect to state and reward functions of the observed variables by Bayesian inference during Policy evaluation. Collected samples modeled as the observed variables and the action value function modeled as the latent variables are used to transform the policy iteration into a problem of inferring the latent variables from the observed ones which were based on the probability theory. In other words, the action value function is being updated according to Bayesian inference at the second stage implementation of a Bayesian reinforcement learning algorithm known as SARSA, in which the action value function as a stochastic variable is being modeled through Gaussian processes at its first stage [12].

5. Conclusion

Bayes' Theorem is extremely successful in providing an intuitively clear alternative method for estimating parameters as well as determining the degree of confidence or uncertainty in these estimates.

Since it is known that (statistical) inference is the process of extracting features about a population or probability distribution from data, it is quite natural that Bayesian inference is defined by making an analogy with the former one as extracting feature about a population or probability distribution from data using Bayes' theorem.

In Bayesian statistics, the information necessary to make inferences is found in observed data, while in frequentist statistics, it is found in other unobserved quantities.

The main advantages of using these tools include the fact that the understanding of complex Bayesian concepts and procedures is achieved in most cases without mathematical formulas and that these tools allow some aspects of Bayesian inference hidden in the mathematical formula of Bayes Theorem to be revealed by the researcher.

Author details

İhsan Ömür Bucak
Iğdır University, Iğdır, Turkey

*Address all correspondence to: iomur.bucak@igdir.edu.tr

References

[1] Daintith J. Bayesian inference. In: A Dictionary of Physics [Internet]. Oxford: OUP Oxford; 2014. Available from: https://www.oxfordreference.com/display/10.1093/oi/authority.20110810104430938 [Accessed: 27 October 2023]

[2] Cheng L, Yin F, Theodoridis S, Chatzis S, Chang TH. Rethinking Bayesian learning for data analysis: The art of prior and inference in sparsity-aware modeling. IEEE Signal Processing Magazine. 2022;**39**(6):18-52. DOI: 10.1109/MSP.2022.3198201

[3] Tipping, ME. Bayesian inference: An introduction to principles and practice in machine learning. In: Bousquet O, von Luxburg U, Rätsch G, editors. Advanced Lectures on Machine Learning. ML 2003. Lecture Notes in Computer Science. Berlin, Heidelberg: Springer; 2004. 3176. DOI: 10.1007/978-3-540-28650-9_3

[4] Purnell DW, Botha EC. Improved performance and generalization of minimum classification error training for continuous speech recognition. In: Proceedings of the Sixth International Conference on Spoken Language Processing, ICSLP 2000/INTERSPEECH 2000. Beijing, China: ISCA; 16-20 Oct 2000. pp. 165-168. DOI: 10.21437/ICSLP.2000-777

[5] Hirsch C, Bharti A, Pedersen T, Waagepetersen R. Bayesian inference for stochastic multipath Radio Channel models. IEEE Transactions on Antennas and Propagation. 2023;**71**(4):3460-3472. DOI: 10.1109/TAP.2022.3215820

[6] Morales J, Yu W. Bayesian inference for neural network based high-precision modeling. In: 2022 IEEE Symposium Series on Computational Intelligence (SSCI). Singapore, Singapore: IEEE; 2022. pp. 442-447. DOI: 10.1109/SSCI51031.2022.10022075

[7] Morales J, Yu W. A novel Bayesian inference based training method for time series forecasting. In: IEEE International Conference on Systems, Man, and Cybernetics (SMC); 2021. Melbourne, Australia: IEEE; 2021. pp. 909-913. DOI: 10.1109/SMC52423.2021.9659009

[8] Torkar R, Furia CA, Feldt R. Bayesian data analysis for software engineering. In: IEEE/ACM 43rd International Conference on Software Engineering: Companion Proceedings (ICSE-Companion); 2021. Madrid, Spain: IEEE; 2021. pp. 328-329. DOI: 10.1109/ICSE-Companion52605.2021.00140

[9] Mhamdi S, Girard P, Virazel A, Bosio A, Ladhar A. Cell-aware diagnosis of customer returns using Bayesian inference. In: 22nd International Symposium on Quality Electronic Design (ISQED); 2021. Santa Clara, CA, USA: IEEE; 2021. pp. 48-53. DOI: 10.1109/ISQED51717.2021.9424337

[10] Shu H. Inference in census-house dataset. In: International Conference on Signal Processing and Machine Learning (CONF-SPML); 2021. Stanford, CA, USA: IEEE; 2021. pp. 282-285. DOI: 10.1109/CONF-SPML54095.2021.00061

[11] Wang YE, Zhu Y, Ko GG, Reagen B, Wei GY, Brooks D. Demystifying Bayesian inference workloads. In: IEEE International Symposium on Performance Analysis of Systems and Software (ISPASS); 2019. Madison,

WI, USA: IEEE; 2019. pp. 177-189. DOI: 10.1109/ISPASS.2019.00031

[12] Xia Z, Zhao D. Online reinforcement learning by Bayesian inference. In: International Joint Conference on Neural Networks (IJCNN); 2015. Killarney, Ireland: IEEE; 2015. pp. 1-6. DOI: 10.1109/IJCNN.2015.7280437

Chapter 2

Indirect Observation of State and Transition Probabilities

Alexei Gilchrist and Lachlan J. Rogers

Abstract

A wide range of systems exhibit stochastic transitions between different states that may be hidden from direct observation. Nevertheless, if the states are coupled to a signal, observation of the signal can provide necessary information to infer the state and switching characteristics. Here we explore a simple hidden Markov model with an observable Poissonian distributed count signal. Determining the parameters of this system from the signal can be difficult in the high-noise regime with non-Bayesian methods. However this system yields a simple Bayesian network description, and variable independencies allow the problem to be formulated in a way that allows tractable inference of the parameters just from the time series. This is an informative demonstration of Bayesian techniques, and in particular the interplay between modelling a system and the process of inference.

Keywords: Bayesian inference, hidden Markov model, state switching, time series, modelling

1. Introduction

One of the really satisfying aspects of Bayesian inference is the structured interplay between modelling and observable data. To begin an analysis, a clear model of the system and noise sources is required. The model informs how to infer parameters from correlations in the observed data. In turn, comparison with experiment might suggest extensions or modifications of the model and lead to more rounds of analysis. We demonstrate this interplay in this chapter with a simple hidden Markov model, that is a distillation of the analysis presented [1] which was applied to quantum emitters. The original system in question had just two states, labelled *on* and *off*, with a probabilistic transition between these states that does not depend on the past of the system. We extend the modelling to an arbitrary number of states and transition probabilities. In the scenario we consider, the system state and transitions are hidden from the observer. What *is* observable is a Poissonian distributed signal whose strength depends on the system state. The task is then to determine the hidden parameters from the observed signal, such as the system state over time, or the switching probabilities.

A variety of systems exhibit stochastic switching between states, sometimes called telegraph noise [2], and the first step to controlling the system might be to understand the characteristics of the switching. An example of such a system is a quantum emitter, where the switching behaviour is called *blinking* due to large intermittent changes in the observed fluorescence. Such emitters are valuable resources for engineered quantum systems, and the presence of blinking limits their applicability. If fact, most quantum light emitters exhibit intermittency in their fluorescence under continuous excitation, including diamond colour centres [3–7], quantum dots [8–12], nanowires [9, 13], nanorods [13], organic semiconductors [14], molecules [15–17], and other systems [18, 19]. Key insights can be distilled from the switching rates and how these depend on relevant parameters, and yet these rates are often the hardest to extract from the raw time series of photon counts in particular for low-intensity light signals or noisy data.

Blinking typically leads to step-like switches in the fluorescence time trace, as illustrated in **Figure 1(a)**. The most common method used to analyse the *on* and *off* states from a blinking time series is threshold analysis [3, 20, 21], illustrated for simulated data in **Figure 1(a)**. However, the choice of threshold intensity is essentially arbitrary, and it can significantly influence the statistics of the *on* and *off* states [22–25]. The threshold technique becomes difficult to use for emitters fluorescing at low signal-to-noise ratios as depicted in **Figure 1(b)**, where the distributions of counts from the *on* and *off* states to overlap. **Figure 1(c)** shows that generalising this situation to three (or more) possible states typically makes it less likely that suitable thresholds can be found to distinguish all of the states.

By taking a Bayesian inference approach we can extract state switching probabilities or rates directly from the time-series of the observable parameter, without resorting to threshold analysis. In order to not over complicate the analysis we will focus on a regime where the state switching rates are slow in comparison to the

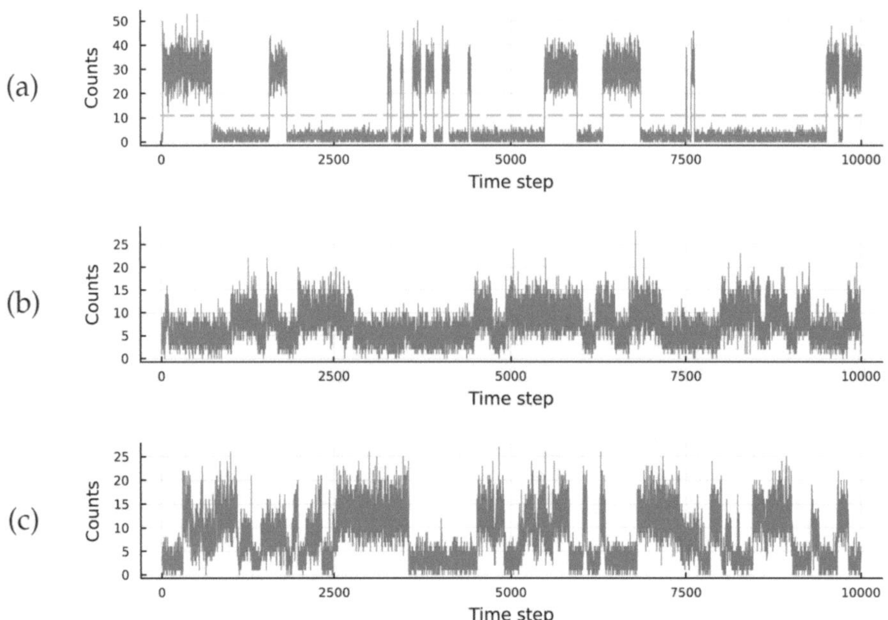

Figure 1.
Not all switching time traces suit analysis in terms of a threshold. (a) Simulated data with clear on and off states. The dotted line represents a threshold, which is used to distinguish the on and off states. (b) An example of the same model with the same switching rates where it would be difficult to do a threshold analysis. (c) A model with three states with no clear distinction.

detection intervals. The analysis can also be extended to the continuous switching regime as demonstrated in [1], either as a sequence of more and more sub-intervals or in the fully continuum limit.

In Section 2 we analyse a switching process where the system is assumed to be either in a number of possible states $\{a, b, c, \ldots\}$ for the whole detection interval. In Section 3 we infer the switching rates from observed data. Finally in Section 4 we use the time trace to infer the system state itself.

2. Modelling

The model for the system has some number of hidden states, $\{a, b, c, \ldots\}$, and it switches randomly from a state x to a state y with probability of T_{xy}. The system state is only indirectly observable by a signal, that is itself Poissonian distributed with a rate R_s that depends on the system state s. The Markov chain for this model is depicted in **Figure 2** for three states.

Such a model can be also used to model a background Poissonian noise process that affects all rates by simply adding the background rate to all other rates, as a combination of Poissonian processes is also Poissonian. In the physical example of quantum emitters, the background rate (known as dark-counts) is due to noise in the detector or electronics. Similarly, while the quantum emitter is in the *on* state and is fluorescing, it is being driven by a strong classical pump so the statistics of the photon counts are also well modelled by a Poissonian process. The *observed* counts are then Poissonian with a rate of either the background rate or the background rate and the fluorescing rate combined, both of which are Poissonian.

We'll write all the state dependent rate parameters as a single variable for brevity in the probability statements $R = R_a R_b R_c \ldots$. The state at the beginning of time-step t

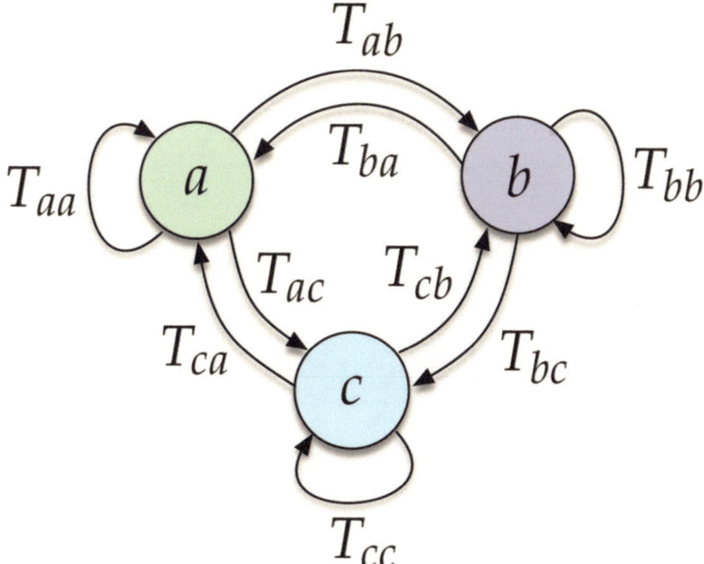

Figure 2.
A discrete time Markov process modelling the evolution of the system between three states. In this case T_{xy} are the switching probabilities from state x to state y. With three states there are potentially six different transition probabilities to track.

(and throughout the time-step) is s_{t-1}, and the probability of seeing c_t counts integrated over the detection window is given by a Poisson distribution with a rate determined by this state,

$$P(c_t|s_{t-1}RI) = c_t^{\lambda} \exp(-\lambda)/c_t! \tag{1}$$

where $\lambda = R_{s_{t-1}}, s_{t-1} \in \{a, b, c, \ldots\}$, and I tags the background information for the model.

Clearly, if we knew the state of the emitter at each data point it would be a simple matter to infer the switching rates. Unfortunately these states are not directly observable, and worse still, because the states are not observed the switching probabilities become dependent on the entire history of the count data. This makes the inference considerably more involved.

We can summarise the dependencies amongst the variables by the Bayesian network (BN) [26] shown in **Figure 3**. The BN can be used to determine conditional independencies between the problem variables, using the property of d-separability. In the BN, s_{t-1} represents the state of the emitter for the tth detector interval. The key variable independencies are the following

$$T \perp R | I \tag{2}$$

$$s_t \perp s_u | s_{t-1} \Omega \quad (u < t-1) \tag{3}$$

$$c_t \perp c_u | s_{t-1} \Omega \quad u \neq t \tag{4}$$

$$c_t \perp s_v | s_{t-1} \Omega \quad v \neq t-1 \tag{5}$$

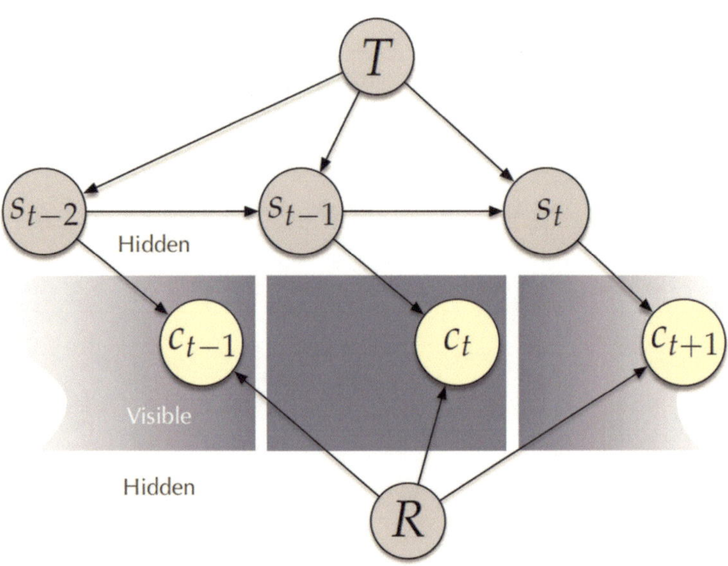

Figure 3.
A partial Bayesian network representing the joint probability distribution of problem parameters. The nodes inside the central dark regions are the visible variables, those outside are hidden parameters. Each dark region represents a detection interval and the visible variables are the accumulated counts for the entire detection interval. We are estimating rate parameters $R_x \in R$, and transition probabilities $T_{xy} \in T$, for $x, y \in \{a, b, c, \ldots\}$ given the observed counts c_t. Enough of the full network is drawn to be able to easily determine the variable independencies.

where $\Omega = TRI$ for compactness. Note that $\sum_y T_{xy} = 1$ for all x, that is all the exit probabilities should add up to one. We will make use of these independencies in the following section.

Generating sample data from the model is straight forward. The initial system state is set at some arbitrary state, say $s_0 = a$. Then at each time step t the next state is chosen according to the probabilities $\{T_{s_t x}\}$ for $x \in \{a, b, c, \ldots\}$. This does a random walk between states following **Figure 2**. Each time step is "observed" by Possonian generated *counts* c_t whose rate is determined by the state: R_{s_t}.

3. Inference

Given the observed counts, the quantity we want to infer is $P\left(T|\vec{c}\,I\right)$ where $\vec{c} \equiv c_1 c_2 \ldots c_N$ is the set of count data obtained over N detector intervals, and T are all the state transition probabilities. Using Bayes' rule, the posterior probability distribution of T can be written as,

$$P\left(T|\vec{c}\,I\right) = \frac{P(T|I)P\left(\vec{c}\,|TI\right)}{P\left(\vec{c}\,|I\right)} \tag{6}$$

The rates of counts R can be added by marginalisation,

$$P\left(T|\vec{c}\,I\right) = \int dR \frac{P(T|I)P\left(\vec{c}\,R|TI\right)}{P\left(\vec{c}\,|I\right)} \tag{7}$$

$$= \frac{1}{P\left(\vec{c}\,|I\right)} \int dR P(TR|I) P\left(\vec{c}\,|\Omega\right). \tag{8}$$

Where the integral is over all the rates $\int dR = \int dR_a \int dR_b \int dR_c \ldots$ for some suitable ranges. Without observing the counts, the independency in (2) implies we can factor $P(TR|I) = P(T|I)P(R|I)$. We will go further and take the probability of all parameters as constant over some initial range (i.e. no prior information) so that

$$P\left(T|\vec{c}\,I\right) = \frac{1}{\mathcal{N}} \int dR P\left(\vec{c}\,|\Omega\right) \tag{9}$$

and determine the normalisation factor $\mathcal{N}$ at the end. Note that $\Omega = TRI$ so these probabilities still depend on the rates and transition probabilities.

If all the states $\vec{s} \equiv s_0 s_1 \ldots s_{N-1}$ were known together with the parameters R and T, the probability of all the counts could again be easily determined. This suggests an approach to the problem by adding these parameters through marginalisation,

$$P\left(\vec{c}\,|\Omega\right) = \sum_{\vec{s}} P\left(\vec{c}\,\vec{s}\,|\Omega\right) = \sum_{\vec{s}} P\left(\vec{c}\,|\vec{s}\,\Omega\right) P\left(\vec{s}\,|\Omega\right) \tag{10}$$

where we abbreviate the summations as $\sum_{\vec{s}} = \sum_{s_0} \ldots \sum_{s_{N-1}}$. Using the independencies (4) and (5), $P\left(\vec{c}\,|\vec{s}\,\Omega\right)$ can be simplified to

$$P\left(\vec{c}\,|\,\vec{s}\,\Omega\right) = \prod_{t=1}^{N} P(c_t|s_{t-1}\Omega), \tag{11}$$

and each term is determined by (1).

The remaining $P\left(\vec{s}\,|\Omega\right)$ term cannot be simply factorised over the states despite being a Markov chain, as observing R and T introduces possible dependencies. We can however expand using the product rule and simplify by making use of the independencies in (3):

$$P\left(\vec{s}\,|\Omega\right) = \prod_{t=1}^{N} P(s_t|s_{t-1}\Omega)P(s_0|\Omega), \tag{12}$$

where we have chosen to expand in temporal order.

Finally, the inference becomes

$$P\left(T|\vec{c}\,I\right) = \frac{1}{\mathcal{N}}\int dR \sum_{\vec{s}} \prod_{t=1}^{N} P(c_t s_t|s_{t-1}\Omega)P(s_0|\Omega) \tag{13}$$

where we have used (5) to write the joint distribution between c_t and s_t.

The problem with (13) is that the sum over $\vec{s}$ contains m^N terms if there are m states, each of which has N products. This will rapidly become intractable as the size of the data grows. Fortunately it's possible to rewrite (13) as a *single* term with N $m \times m$ matrix products. Consider the following matrix written for a three-state system:

$$Q_t = \begin{bmatrix} P(c_t s_t = a|s_{t-1} = a\,\Omega) & P(c_t s_t = a|s_{t-1} = b\,\Omega) & P(c_t s_t = a|s_{t-1} = c\,\Omega) \\ P(c_t s_t = b|s_{t-1} = a\,\Omega) & P(c_t s_t = b|s_{t-1} = b\,\Omega) & P(c_t s_t = b|s_{t-1} = c\,\Omega) \\ P(c_t s_t = c|s_{t-1} = a\,\Omega) & P(c_t s_t = c|s_{t-1} = b\,\Omega) & P(c_t s_t = c|s_{t-1} = c\,\Omega) \end{bmatrix}, \tag{14}$$

and vector

$$\vec{D}_0 = \begin{bmatrix} P(s_0 = a|\Omega) \\ P(s_0 = b|\Omega) \\ P(s_0 = c|\Omega) \end{bmatrix} \tag{15}$$

then

$$\sum_{\vec{s}} \prod_{t=1}^{N} P(c_t s_t|s_{t-1}\Omega)P(s_0|\Omega) = \vec{1}^{\,T} \prod_{t=1}^{N} Q_t \vec{D}_0 \tag{16}$$

where the product on the right hand side is read as decreasing in t to the right. The equivalence of (16) to (13) is readily verified by expanding a few terms out. The matrix multiplication will sum over the columns of Q_t which is a summation over the states s_{t-1}. Each successive multiplication sums over another state and the final multiplication by the row vector $\vec{1}^{\,T}$ (for the three state system this is $[1\ 1\ 1]$) will sum over s_N. The generalisation of Q_t and $\vec{D}_0$ to more than three states is straight forward. With this equivalence, the inference reads

$$P\left(T|\vec{c}\,I\right) = \frac{1}{\mathcal{N}}\int dR\,\vec{1}^{T}\prod_{t=1}^{N} Q_t\vec{D}_0. \tag{17}$$

There is one outstanding issue in calculating (17), and that is that care must be taken to avoid underflow or overflow. The normalisation factor includes $P\left(\vec{c}\,|I\right)$ which is the probability of the observed counts given no knowledge of the parameters or transition probabilities. While it could be expanded by adding parameters and states it would be difficult to calculate. Without it, the computation of (17) quickly leads to overflow. The key to controlling this normalisation issue is to observe that (17) is similar to a time evolving state

$$\prod_{t=1}^{N} Q_t\vec{D}_0 = Q_N \ldots Q_3Q_2Q_1\vec{D}_0 = Q_N \ldots Q_3Q_2\vec{D}_1 = Q_N \ldots Q_3\vec{D}_2. \tag{18}$$

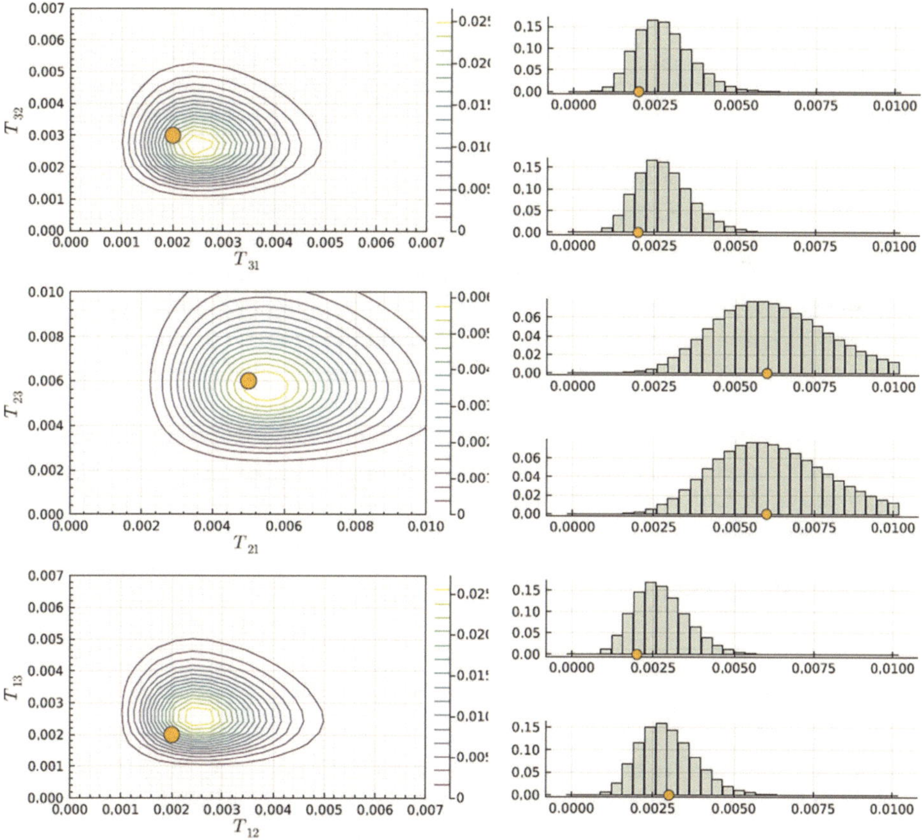

Figure 4.
*Inference of the switching probabilities on simulated data given in **Figure 1(c)**. The true value is shown as a dot in each plot. (Left) Joint marginal probability distributions for the exit probabilities for each state. (Right) Marginal probability distribution for each separate exit probability.*

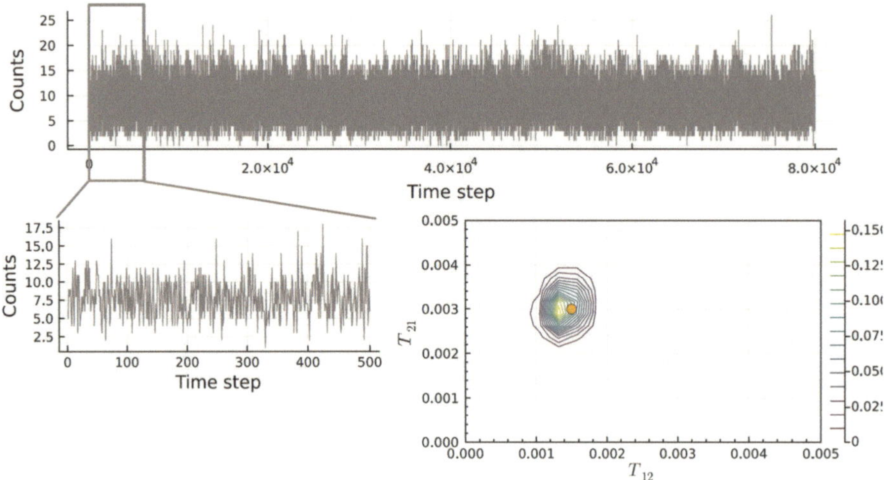

Figure 5.
Time series with high relatively switching rates an difficult to discern states. Nevertheless the constraints imposed by the model allow the switching rate to be accurately estimated with enough data.

So we can compute the "state" D_t for a multi-dimensional grid of parameter values that spans the inference space, and at each time step re-normalise since the sum of the probabilities over the entire grid should be 1. In this way the normalisation can be controlled.

As a demonstration of the algorithm, **Figure 4** shows the marginals of the posterior distributions for all the exit probabilities from each state inferred from the data of **Figure 1(c)**. The count rates R_a, R_b and R_c where assumed known for the purpose of the simulation. It should be noted that the inference contains no approximations or arbitrary choices, and makes use of all the data. The key point of contention with observed data is whether the model is realistic to the system. If it is, then the inference faithfully converges on the underlying values regardless of what they are. If it is not, then this forms the point of departure for another round of modelling.

To illustrate the point, consider a model that is transitioning quickly between two states that are not very distinguishable. See for example the time series in **Figure 5**. It would be difficult to discern that there are two levels, let alone to estimate their switching rates. The model, and the assumption that there exists only two states, means that the inference will weigh each data point against the model and allow the extraction of the switching rates despite the noise.

4. Inferring the state

A further inference we can make is to determine the underlying state behind a data point given all the data observed, but without knowledge of the transition probabilities T_{xy} or the count rates R_x. Specifically, the task is to determine $P\left(s_k = x|\vec{c}\, I\right)$, where $x \in \{a, b, c, \dots\}$.

In order to do this inference we add T, R and the rest of the states using marginalisation. An application of Bayes' rule then yields,

$$P\left(s_k = x|\vec{c}\,I\right) = \frac{1}{\mathcal{N}}\int dTdR\sum_{\vec{s}\neq s_x} P\left(\vec{c}\,|s_1 \dots s_{k-1}\left(s_k = x\right)s_{k+1}\dots s_n\Omega\right)$$
$$\times P(s_1 \dots s_{k-1}\left(s_k = x\right)s_{k+1}\dots s_n|\Omega) \tag{19}$$

where $\Omega = TRI$ as before, and the prior probability distributions where taken as constant with the normalisation $\mathcal{N}$ to be determined at the end. The summation is over all states s_j where $j \neq k$ denoted in short as $\vec{s} \neq s_k$. Expanding the last term in temporal order and making use of the independencies in (3) and (4) leads to,

$$P\left(s_k = x|\vec{c}\,I\right) = \frac{1}{\mathcal{N}}\int dRdT\sum_{\vec{s}\neq s_k}\prod_{t=1}^{N} P(c_t s_t|s_{t-1}\Omega)P(s_0|\Omega). \tag{20}$$

The probability can again be efficiently computed using the matrices Q_t (14) and vector $\vec{D}_0$ (15) but with a restricted choice for Q_k,

$$P\left(s_k = x|\vec{c}\,I\right) = \frac{1}{\mathcal{N}}\int dRdT\ \vec{1}^T Q_n \cdots Q_{k+1} Q_k^{(x)} Q_{k-1} \cdots Q_1 \vec{D}_0. \tag{21}$$

where $Q_k^{(x)}$ only has the row corresponding to state x. So for the three-state example we would have

$$Q_k^{(b)} = \begin{bmatrix} 0 & 0 & 0 \\ P(c_k s_k = b|s_{k-1} = a\Omega) & P(c_k s_k = b|s_{k-1} = b\Omega) & P(c_k s_k = b|s_{k-1} = c\Omega) \\ 0 & 0 & 0 \end{bmatrix}. \tag{22}$$

5. Conclusions

We have derived efficient methods for obtaining the posterior distribution of transition rates given observation of accumulated signal for a hidden Markov model with multiple states. The methods do not require approximations to make the calculation tractable, and make use of all the observed data; the main fitting task is the selection and assumptions that go into the models. Moreover, the advantage of obtaining the posterior distribution is that rigorous error bounds can be placed on the inferred parameters for example by use of credible regions. We have also demonstrated how to use the data to determine the underlying state, and unlike the threshold technique, the result automatically carries error bounds in the form of a probability distribution. In general we expect the methods presented to apply to a wide variety of hidden Markov models with multiple states and other methods of indirect observation. The treatment here has looked at systems where there is a well-defined state time, which will apply to situations where the signal is sampled faster than the typical transition rates. These models are then capable of determining the model parameters for high or low switching probabilities. It is possible to generalise this approach to model systems which might switch at any time regardless of signal acquisition intervals [1].

Acknowledgements

This work was supported by the Australian Research Council (ARC) Centre of Excellence for Quantum Engineered Systems grant (CE170100009).

We recognise the intellectual and physical labour of this research was conducted on the traditional lands of the Wattamattagal clan of the Darug nation, and of the Pambalong clan of the Awabakal people.

Author details

Alexei Gilchrist[1,2*] and Lachlan J. Rogers[2,3]

1 Department of Physics and Astronomy, Macquarie University, Sydney, NSW, Australia

2 ARC Centre of Excellence for Engineered Quantum Systems, Australia

3 School of Information and Physical Sciences, University of Newcastle, Newcastle, NSW, Australia

*Address all correspondence to: alexei@entropy.energy

References

[1] Geordy J, Rogers LJ, Rogers CM, Volz T, Gilchrist A. Bayesian estimation of switching rates for blinking emitters. New Journal of Physics. 2019;**21**(6):063001. DOI: 10.1088/1367-2630/ab1dfd

[2] Efros AL, Rosen M. Random telegraph signal in the photoluminescence intensity of a single quantum dot. Physical Review Letters. 1997;**78**:1110-1113. DOI: 10.1103/PhysRevLett.78.1110

[3] Jantzen U, Kurz AB, Rudnicki DS, Schäfermeier C, Jahnke KD, Andersen UL, et al. Nanodiamonds carrying silicon-vacancy quantum emitters with almost lifetime-limited linewidths. New Journal of Physics. 2016;**18**(7):073036. Available from: http://stacks.iop.org/1367-2630/18/i=7/a=073036

[4] Neu E, Agio M, Becher C. Photophysics of single silicon vacancy centers in diamond: Implications for single photon emission. Optics Express. 2012;**20**(18):19956-19971

[5] Bradac C, Gaebel T, Naidoo N, Sellars M, Twamley J, Brown L, et al. Observation and control of blinking nitrogen-vacancy centres in discrete nanodiamonds. Nature Nanotechnology. 2010;**5**(5):345

[6] Castelletto S, Edmonds A. 680-890 nm spectral range of nickel-nitrogen and nickel-silicon complex single centres in diamond. In: Quantum Communications and Quantum Imaging X. Vol. 8518. San Diego, California, United States: International Society for Optics and Photonics; 2012. p. 85180R. DOI: 10.1117/12.958606

[7] Wang C. A Solid-state Single Photon Source Based on Color Centers in Diamond. Munich, Germany: LMU; 2007. DOI: 10.1109/CLEOE-IQEC.2007.4386885

[8] Nirmal M, Dabbousi BO, Bawendi MG, Macklin J, Trautman J, Harris T, et al. Fluorescence intermittency in single cadmium selenide nanocrystals. Nature. 1996;**383**(6603):802

[9] Frantsuzov P, Kuno M, Janko B, Marcus RA. Universal emission intermittency in quantum dots, nanorods and nanowires. Nature Physics. 2008;**4**(7):519

[10] Efros AL, Nesbitt DJ. Origin and control of blinking in quantum dots. Nature Nanotechnology. 2016;**11**(8):661

[11] Galland C, Ghosh Y, Steinbrück A, Sykora M, Hollingsworth JA, Klimov VI, et al. Two types of luminescence blinking revealed by spectroelectrochemistry of single quantum dots. Nature. 2011; **479**(7372):203

[12] Galland C, Ghosh Y, Steinbrück A, Hollingsworth JA, Htoon H, Klimov VI. Lifetime blinking in nonblinking nanocrystal quantum dots. Nature Communications. 2012;**3**:908

[13] Frantsuzov PA, Volkán-Kacsó S, Jankó B. Universality of the fluorescence intermittency in nanoscale systems: Experiment and theory. Nano Letters. 2013;**13**(2):402-408

[14] Ruth A, Hayashi M, Zapol P, Si J, McDonald MP, Morozov YV, et al. Fluorescence intermittency originates from reclustering in two-dimensional organic semiconductors. Nature Communications. 2017;**8**:14521

[15] Bout DAV, Yip WT, Hu D, Fu DK, Swager TM, Barbara PF. Discrete

intensity jumps and intramolecular electronic energy transfer in the spectroscopy of single conjugated polymer molecules. Science. 1997; **277**(5329):1074-1077. Available from: https://science.sciencemag.org/content/277/5329/1074

[16] Dickson RM, Cubitt AB, Tsien RY, Moerner WE. On/off blinking and switching behaviour of single molecules of green fluorescent protein. Nature. 1997;**388**:355-358

[17] Haase M, Hübner CG, Reuther E, Herrmann A, Müllen K, Basché T. Exponential and power-law kinetics in single-molecule fluorescence intermittency. The Journal of Physical Chemistry B. 2004;**108**(29): 10445-10450. DOI: 10.1021/jp0313674

[18] Berhane AM, Bradac C, Aharonovich I. Photoinduced blinking in a solid-state quantum system. Physical Review B. 2017;**96**(4):041203

[19] Gammelmark S, Mølmer K, Alt W, Kampschulte T, Meschede D. Hidden Markov model of atomic quantum jump dynamics in an optically probed cavity. Physical Review A. 2014;**89**(4):043839

[20] Kuno M, Fromm DP, Hamann HF, Gallagher A, Nesbitt DJ. Nonexponential "blinking" kinetics of single CdSe quantum dots: A universal power law behavior. The Journal of Chemical Physics. 2000;**112**(7):3117-3120

[21] Kuno M, Fromm D, Hamann H, Gallagher A, Nesbitt DJ. "On"/"off" fluorescence intermittency of single semiconductor quantum dots. The Journal of Chemical Physics. 2001; **115**(2):1028-1040

[22] Crouch CH, Sauter O, Wu X, Purcell R, Querner C, Drndic M, et al. Facts and artifacts in the blinking statistics of semiconductor nanocrystals. Nano Letters. 2010;**10**(5):1692-1698

[23] Frantsuzov PA, Volkán-Kacsó S, Jankó B. Model of fluorescence intermittency of single colloidal semiconductor quantum dots using multiple recombination centers. Physical Review Letters. 2009;**103**(20):207402

[24] Watkins LP, Yang H. Detection of intensity change points in time-resolved single-molecule measurements. The Journal of Physical Chemistry B. 2005; **109**(1):617-628

[25] Amecke N, Heber A, Cichos F. Distortion of power law blinking with binning and thresholding. The Journal of Chemical Physics. 2014;**140**(11):114306

[26] Darwiche A. Modeling and Reasoning with Bayesian Networks. Cambridge: Cambridge University Press; 2009. DOI: 10.1017/CBO9780511811357

Chapter 3

Nested Sampling: A Case Study in Parameter Estimation

Fesih Keskin

Abstract

Nested Sampling (NS) is a powerful Bayesian inference algorithm that can be used to estimate parameter posteriors and marginal likelihoods for complex models. It is a sequential algorithm that works by iteratively removing low-likelihood regions of the parameter space while keeping track of the weights of the remaining points. This allows NS to efficiently sample the posterior distribution, even for models with complex and multimodal posteriors. NS has been used to estimate parameters in a wide range of applications, including cosmology, astrophysics, biology, and machine learning. It is particularly well-suited for problems where the posterior distribution is difficult to sample directly or where it is important to obtain accurate estimates of the marginal likelihood. This study explores the potential of NS as an alternative to these traditional methods for Direction of Arrival (DoA) estimation. Capitalizing on its strengths in handling multimodal distributions and dimensionality, we explore its applicability, practical application, and comparative performance. Through a simulated case study, we demonstrate the potential superiority of NS in certain challenging conditions. However, it also exposes computational intensity and forced antecedent selection as challenges. In navigating this exploration, we provide insights that advocate for the continued investigation and development of NS in broader signal-processing settings.

Keywords: nested sampling (NS), Bayesian inference, posterior distribution, marginal likelihood, direction of arrival (DoA) estimation

1. Introduction

Nested Sampling (NS) is a Bayesian inference algorithm that has gained significant popularity across multiple scientific fields due to its ability to accurately estimate parameter posteriors and marginal likelihoods, particularly in models with complex and multimodal posteriors [1]. NS is a sequential algorithm that continually refines the parameter space, concentrating on high-likelihood regions, effectively capturing the posterior distribution [2]. Its versatility in addressing challenges where direct sampling of the posterior distribution is difficult or where precise marginal likelihood estimates are crucial makes it a valuable tool in a range of fields, including cosmology, astrophysics, biology, and machine learning [3, 4].

The ability of NS to accurately estimate parameters in complex, high-dimensional models makes it a valuable tool for array signal-processing applications [5]. Sensor arrays have a wide range of applications in fields including radar, sonar, wireless communications, and seismology [6]. They offer spatial selectivity, which can be utilized to estimate the Direction of Arrival (DoA) of impinging signals [6]. While traditional methods for DoA estimation, such as the MUSIC (Multiple Signal Classification) [7] and ESPRIT (Estimation of Signal Parameters via Rotational Invariance Techniques) [8] algorithms, have their limitations, the NS approach may offer advantages in certain scenarios. For example, estimating the DoA may be challenging, as some methods necessitate knowing the number of sources ahead of time, while others may struggle with model mismatches or coherent signals [6]. NS presents a hopeful alternative in DoA estimation, given its ability to efficiently explore the parameter space.

The application of NS in this scenario is pivotal, as it allows for a more nuanced approach to DoA estimation. The algorithm's ability to efficiently traverse through the parameter space by focusing on regions with higher likelihood scores is particularly beneficial when dealing with multiple source signals that may present multimodal characteristics in their distribution [9]. This is especially relevant in complex acoustic or electromagnetic environments where sources may have varied and overlapping signal properties. This study aims to provide comprehensive insights into the applicability and practical application of NS in array signal processing, particularly in DoA estimation. While highlighting the strengths of NS in dealing with complex signal-processing scenarios, we acknowledge and address its limitations. Moreover, this study presents NS not only as an algorithm but also as a transformative approach in the field of signal processing, especially in DoA estimation.

This study is organized into five sections. Section 2 reviews the existing literature on NS and its applications, particularly in parameter estimation. In Section 3 we describe the signal model and application of NS. Section 4 presents simulation results, showing effectiveness of the NS. Finally, Section 5 concludes the study.

2. Related works

NS, a significant advancement in Bayesian inference, was introduced by Skilling [1]. This method revolutionized the approach to correlating likelihood functions with prior mass, showcasing its capability to efficiently obtain evidence while managing complex Bayesian problems. Skilling's work laid the foundation for NS's independence from actual likelihood values and its ability to address phase-change issues, setting it apart from traditional methods like annealing [1]. Building on these concepts, Aitken and Akman applied NS in systems biology, particularly in analyzing circadian rhythm models. Their research demonstrated the impact of time series length on posterior parameter densities, offering insights into the constraints imposed by increasing circadian cycle data. They also highlighted the utility of the coefficient of variation in distinguishing between parameters with varying degrees of constraint [10].

Veitch et al. underscored the effectiveness of NS algorithms in the field of gravitational-wave observations from compact binary coalescences. Their work with the LALInference software library highlighted the precision and robustness of NS in parameter estimation, marking a significant contribution to this area of astrophysics [4]. Further exploring NS's capabilities, Feroz et al. focused on advancements in the MultiNest algorithm, particularly the implementation of importance NS (INS). This development was crucial in enhancing Bayesian evidence calculation accuracy in fields such as astrophysics, cosmology, and particle physics without altering the parameter space exploration by MultiNest [9]. Speagle introduced DYNESTY, a dynamic NS Python package designed for estimating Bayesian posteriors and evidence in astronomical analyses. This package, building on previous work, demonstrated significant improvements in sampling efficiency, particularly in complex, multimodal distributions [11]. Buchner expanded the scope of NS through a comprehensive review, discussing its proficiency in navigating complex, potentially multimodal posteriors. This review not only addresses various NS variants but also delves into the practical implementation challenges and recent advancements in diagnostics and visualization [12].

NS has found diverse applications across various fields. Tawara et al. proposed a novel nested Gibbs sampling method for estimating mixture-of-mixture models, particularly effective in complex multilevel data like speech utterances. This method showcased its efficiency in avoiding local optima and outperforming conventional methods, indicating its potential for application in areas like image clustering [13]. Buchner et al. introduced a scalable algorithm for analyzing massive datasets with complex physical models. Their approach, collaborative NS, significantly reduced the number of model evaluations required, adeptly managing heterogeneous data and noise properties, making it particularly effective for large-scale astronomical surveys and simulations [14]. Alsing and Handley addressed a key limitation in NS implementations related to the specification of priors. They proposed a novel approach using parametric bijectors trained on samples from target priors, facilitating the use of NS under arbitrary priors. This method was particularly relevant for cases where posterior distributions from one experiment served as priors for another [15]. Ashton et al. provided a comprehensive review of Skilling's NS algorithm. They emphasized its role in Bayesian inference and multi-dimensional integration across various scientific fields including cosmology, gravitational-wave astronomy, and materials science. The review highlighted the algorithm's adaptability and potential for future developments, particularly in relation to machine learning and high-dimensional data challenges [16].

In the realm of parameter estimation, Vitale et al. investigated the impact of calibration errors on Bayesian parameter estimation for gravitational wave signals from inspiral binary systems. Their use of NS revealed that calibration errors introduced systematic shifts in parameter estimates that were relatively small compared to statistical measurement errors, ensuring the reliability of parameter estimation even with potential calibration inaccuracies [17]. Allison and Dunkley compared three sampling methods—Metropolis-Hasting, NS, and affine-invariant ensemble Markov chain Monte Carlo—for Bayesian parameter estimation. They found that NS was particularly effective in delivering accurate posterior statistics at a lower computational cost, favoring it over Metropolis-Hasting in many instances [18]. Higson et al. identified two primary sources of sampling errors in NS parameter estimation and proposed a novel algorithm for estimating these errors. Their method, which could be integrated into existing NS software, was empirically validated for accuracy, addressing a crucial gap in NS parameter estimation [19]. Thrane and Talbot provided an accessible introduction to Bayesian inference, focusing on hierarchical models and hyper-parameters, using examples primarily from gravitational-wave astronomy. Their work covered the basics of Bayesian analysis, including likelihoods, priors, posteriors, and sampling methods, and extended to more advanced topics like hyper-parameters and hierarchical models [20]. Higson et al. presented dynamic NS, an improvement over standard NS that varied the number of live points for more efficient allocation of posterior samples. This method demonstrated significant increases in computational efficiency for parameter estimation and evidence calculations, validating its effectiveness across various likelihoods, priors, and dimensions [5]. Paulen et al. addressed the challenge of identifying feasible parameter sets in nonlinear dynamic systems. They introduced an innovative NS technique from Bayesian statistics, proving effective in inner-approximating these parameter sets. Validated by various case studies, this approach offered a promising alternative to existing methods and suggested a new direction for future research combining NS with set-theoretic concepts [21].

Jasa and Xiang applied the NS algorithm in Bayesian room-acoustics decay analysis, demonstrating its effectiveness in discriminating the number of energy decays in acoustically coupled spaces. Their work highlighted NS's capability in handling two levels of Bayesian inference, decay order estimation and decay parameter estimation, offering significant insights and potential improvements in architectural acoustics applications [22]. Fackler, Xiang, and Horoshenkov utilized Bayesian probabilistic inference and NS to analyze the pore microstructure of multilayer porous media in acoustical applications. Their method effectively determined the number of layers and their physical properties in porous specimens, providing a quantitative approach for model selection and enhancing understanding of the layers' characteristics through the Bayesian evidence and posterior distribution analysis [23]. Landschoot and Xiang applied a model-based Bayesian approach using NS for the DoA analysis of sound sources, employing a spherical microphone array. Their method effectively estimated the number of sound sources and their DoAs, demonstrating improvements over previous techniques and suggesting potential for broader applications in room acoustics and complex noise environments [24].

3. Methodology

This section provides a detailed overview of our approach to applying NS in DoA estimation, encompassing the signal model setup, exploration of the likelihood function, discussion of prior distributions, and initialization of NS parameters. Within this study, we denote vectors and matrices using bold lowercase and uppercase letters, respectively. The transpose of a matrix is indicated by $[\cdot]^T$. Furthermore, $\mathbf{I}_M$ is used to represent an *MxM* identity matrix.

3.1 Signal model

Consider an array composed of M identical and omnidirectional sensors placed at positions $\mathbf{p}_m = [x_m,\ y_m,\ z_m]^T$ for $m = 1, \dots, M$. The array receives L narrow-band, far-field source signals from distinct directions $\Theta_l = (\phi_l,\ \theta_l)$ for $l = 1, \dots, L$. The azimuth angle (ϕ) measured counter-clockwise from the positive x-axis on the x-y plane, ranging from 0° to 180°, while the elevation (θ) is the clockwise angle from the positive z-axis, 0° to 90°, as shown in **Figure 1**.

The *Mx1* array response $\mathbf{y}(t)$ at time t can be described as [25],

$$\mathbf{y}(t) = \mathbf{A}(\Theta) \cdot \mathbf{s}(t) + \mathbf{n}(t), \quad t = 1, \dots, T, \tag{1}$$

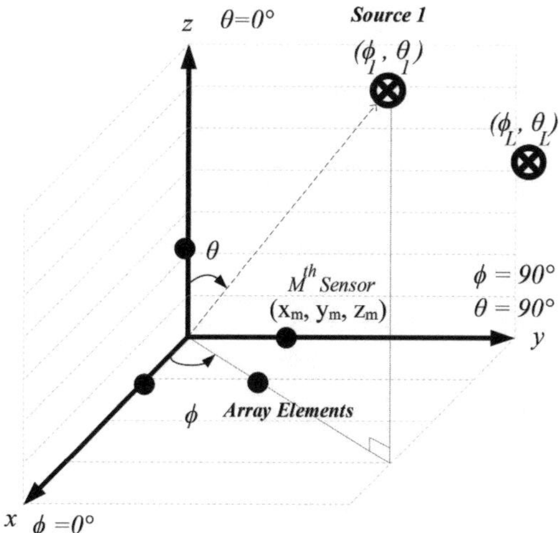

Figure 1.
Three-dimensional (3-D) coordinate system illustrating the configuration of an array for DoA estimation of multiple source signals.

where T is the number of observations. The source signal $\mathbf{s}(t)$ is an $Lx1$ vector, and the noise $\mathbf{n}(t)$ is an $Mx1$ vector, both assumed to be uncorrelated, zero-mean, and spatially and temporally white Gaussian complex random processes. $\mathbf{A}$ is the steering matrix, whose columns are the steering vectors $\mathbf{a}(\Theta_l)$ corresponding to each source direction as follows,

$$\mathbf{a}(\Theta_l) = [a_1(\Theta_l),\ a_2(\Theta_l),\ \cdots,\ a_M(\Theta_l)]^T, \tag{2}$$

where,

$$a_m(\Theta_l) = e^{-j\frac{2\pi}{\lambda}\left(\mathbf{p}_m^T \cdot \mathbf{u}_l\right)}, \tag{3}$$

where $\mathbf{u}_l = [\sin(\theta_l)\cos(\phi_l),\ \sin(\theta_l)\sin(\phi_l),\ \cos(\theta_l)]^T$ is the unit direction vector of the lth source signal with the wavelength of λ.

3.2 NS for DoA estimation

NS is a Bayesian inference technique that is particularly well-suited for problems involving the estimation of model parameters. The goal is to compute the posterior distribution of the parameters of interest, given the observed data. For the DoA estimation problem, our goal is to estimate the directions $(\phi_l,\ \theta_l)$ from the observed data given in (1).

3.2.1 Likelihood function

The first step in applying NS is to define the likelihood function, which quantifies how well our model predicts the observed data for a given set of parameters. Since the noise $\mathbf{n}(t)$ is assumed to be a complex Gaussian random variable with zero mean and variance σ_n^2,

$$n_m(t) \sim \mathcal{CN}(0, \sigma_n^2). \tag{4}$$

For an M element sensor array, the joint pdf conditioned on Θ for T snapshot can be written as [6],

$$\mathcal{L}(\Theta) = p(\mathbf{y}(t)|\mathbf{s}(t), \Theta) = \prod_{t=1}^{T} \frac{1}{\sqrt{2\pi\sigma_n^2}} \exp\left(-\frac{|\mathbf{y}(t) - \mathbf{A}(\Theta) \cdot \mathbf{s}(t)|^2}{2\sigma_n^2} \right), \tag{5}$$

where Θ represents the set of parameters to be estimated, that is, the directions $(\phi_l,\ \theta_l)$, and σ_n^2 is the variance of the noise. Taking the natural logarithm of the both sides of (5), the log-likelihood is [6],

$$\ln p(\mathbf{y}(t)|\mathbf{s}(t), \Theta) = -\frac{MT}{2} \ln(2\pi\sigma_n^2) - \frac{1}{2\sigma_n^2} \sum_{t=1}^{T} |\mathbf{y}(t) - \mathbf{A}(\Theta) \cdot \mathbf{s}(t)|^2. \tag{6}$$

3.2.2 Prior distribution

Next, we need to define a prior distribution, $\boldsymbol{\pi}$, over the parameters Θ. This encodes our prior beliefs about the parameters before observing the data. A common starting point, when no specific prior information is available, is to use a uniform prior. This non-informative prior expresses equal uncertainty over all possible directions from which signals could arrive. For azimuth and elevation, the uniform priors are defined over their respective feasible intervals.

The prior for the azimuth angle ϕ is given by,

$$\boldsymbol{\pi}(\phi_l) = \begin{cases} \frac{1}{\pi} & \text{if } 0 \leq \phi < \pi, \\ 0 & \text{otherwise.} \end{cases} \tag{7}$$

The prior for the elevation angle θ is given by,

$$\boldsymbol{\pi}(\theta_l) = \begin{cases} \frac{2}{\pi} & \text{if } 0 \leq \theta < \frac{\pi}{2}, \\ 0 & \text{otherwise.} \end{cases} \tag{8}$$

These priors ensure that all possible DoAs are treated equally before observing the data. Furthermore, they represent our belief about the parameters before observing any data. They can be updated with the likelihood function to obtain the posterior distribution, which represents our updated belief about the parameters after observing the data. Note that if we have prior knowledge indicating a likelihood of specific directions, such as broadside directions in Linear Arrays, it's common to use Gaussian priors. These are designed around mean directions μ_ϕ and μ_θ with accompanying standard deviations $(\sigma_\phi, \sigma_\theta)$ that express our confidence level in these averages.

3.2.3 Bayesian evidence

Bayesian Evidence, $\mathcal{Z}$, serves as a critical metric in model fitting and comparison within the Bayesian framework. It quantifies how well a model fits the data and is defined as the integral of the likelihood function across the parameter space, weighted by the prior distribution as follows,

$$\mathcal{Z} = \int \mathcal{L}(\Theta) \times \pi(\Theta) d\Theta. \tag{9}$$

In high-dimensional spaces, direct computation of $\mathcal{Z}$ is complex, but NS provides an effective estimation method. The value of $\mathcal{Z}$ is pivotal for model selection, where higher evidence indicates a more preferable model in terms of data explanation and model complexity.

3.2.4 Posterior distribution

Following the computation of Bayesian evidence, the focus shifts to the posterior distribution, which is central to Bayesian inference, especially in DoA estimation. The posterior is derived using Bayes' theorem as:

$$p(\Theta|\text{data}) = \frac{\mathcal{L}(\Theta) \times \pi(\Theta)}{\mathcal{Z}}. \tag{10}$$

Here, $\mathcal{L}(\Theta)$ represents the likelihood function (5), and $\pi(\Theta)$ denotes the prior distribution (8). This formulation updates our knowledge about the parameters Θ after considering the observed data. NS's role is to effectively traverse this often complex, multidimensional posterior distribution.

3.2.5 Initialization of NS

The initialization stage of the NS algorithm for DoA estimation begins with the generation of live points from the prior distribution. We define the complete set of DoA parameters for all sources as $\Theta = \{\Theta_1, \Theta_2, \dots, \Theta_L\}$.

Each parameter within Θ_l is assumed to have a prior distribution $\boldsymbol{\pi}(\Theta_l)$, which is often chosen to be uniform over the physically meaningful range for both azimuth ϕ_l and elevation θ_l as given in (7) and (8), respectively.

For the jth live point, the parameters are sampled according to the prior distributions,

$$\Theta^{(j)} = \left\{\Theta_1^{(j)}, \Theta_2^{(j)}, \dots, \Theta_L^{(j)}\right\}, \tag{11}$$

where,

$$\Theta_l^{(j)} = \left(\phi_l^{(j)}, \theta_l^{(j)}\right) \sim \pi(\phi_l) \times \pi(\theta_l), \quad l = 1, \dots, L. \tag{12}$$

Each live point $\Theta^{(j)}$ corresponds to a hypothesis about the directions of the L source signals. The likelihood $\mathcal{L}(\Theta^{(j)})$ for each live point is computed based on the array response to these hypothesized directions, and the far-field assumption simplifies the array response matrix $\mathbf{A}(\Theta^{(j)})$. With the live points initialized, the NS algorithm proceeds to iterate, successively refining these points toward the higher likelihood regions of the parameter space, thus exploring the posterior distribution of the DoA parameters.

3.2.6 Sampling loop

In each iteration, the point with the lowest likelihood (the "worst" point) from the set of live points is identified and removed as,

$$\mathcal{L}_{\text{worst}}^{(k)} = \min\left\{\mathcal{L}\left(\Theta^{(1)}\right), \dots, \mathcal{L}\left(\Theta^{(N)}\right)\right\} \tag{13}$$

where $\mathcal{L}\left(\Theta^{(j)}\right)$ is the likelihood for the jth live point out of N total points. The parameters of the worst point, $\Theta_{\text{worst}}^{(k)}$, are recorded, and then, the point is discarded. After identifying and removing the worst live point with the lowest likelihood, L_{worst}, the next step is to sample a new point from the prior distribution with a likelihood greater than L_{worst} as follows,

$$\Theta_{\text{new}} \sim \pi(\Theta | \mathcal{L}(\Theta) > L_{\text{worst}}), \tag{14}$$

where Θ_{new} is the new point in the parameter space, and $\pi(\Theta | \mathcal{L}(\Theta) > L_{\text{worst}})$ represents the constrained prior distribution. The condition $\mathcal{L}(\Theta) > L_{\text{worst}}$ ensures that the new point has a higher likelihood than the worst point that was just removed. This process iterates, successively concentrating the live points in regions of higher likelihood, until the algorithm's stopping criteria are reached.

3.2.7 Evidence calculation

The estimation of the Bayesian evidence, denoted as $\mathcal{Z}$, is updated iteratively with the removal of each worst live point. At iteration k, the contribution to the evidence from the discarded point $\Theta_{\text{worst}}^{(k)}$ is given by,

$$\mathcal{Z}_k = \mathcal{L}_{\text{worst}}^{(k)} \times w_k, \tag{15}$$

where $\mathcal{L}_{\text{worst}}^{(k)}$ is the likelihood of the worst point at iteration k, and w_k is the weight representing the "prior volume" covered by the point. This prior volume fraction is typically computed as $w_k = \frac{1}{2}(\mathcal{X}_{k-1} - \mathcal{X}_{k+1})$, where $\mathcal{X}_k$ is the cumulative prior mass contained within the set of live points at iteration k.

The total evidence after K iterations is then approximated by summing over all such contributions,

$$\mathcal{Z} \approx \sum_{k=1}^{K} \mathcal{Z}_k. \tag{16}$$

The prior mass $\mathcal{X}_k$ is assumed to decrease exponentially with each iteration, reflecting the NS process where the live points are iteratively confined to regions of higher likelihood. The process terminates when the expected increase in evidence due to the remaining live points falls below a certain threshold as,

$$\Delta\mathcal{Z}_k = \mathcal{L}_{\max}^{(k)} \times \mathcal{X}_k < \varepsilon \times \mathcal{Z}, \tag{17}$$

where $\mathcal{L}_{\max}^{(k)}$ is the maximum likelihood among the remaining live points, $\mathcal{X}_k$ represents the fraction of the prior volume corresponding to the live points, and ε is a small positive threshold value. The algorithm concludes when $\Delta\mathcal{Z}_k$ is sufficiently small, indicating that further iterations are unlikely to significantly alter the evidence $\mathcal{Z}$.

As summarized in Algorithm 1, NS is applied to the DoA estimation problem. The initialization process begins with generating a set of live points from the prior distributions, which represent hypotheses about the source directions. The iterative sampling loop continues to refine these points based on their likelihoods until the convergence criteria are met, ultimately yielding estimates for the DoA parameters and the Bayesian evidence.

Algorithm 1 NS for DoA Estimation

Require: Array response $\mathbf{y}$.
Ensure: DoA Parameter Estimates Θ, Bayesian Evidence $\mathcal{Z}$.
Initialization:
Generate N live points $\left\{\Theta^{(1)}, \dots, \Theta^{(N)}\right\}$ from the prior $\pi(\Theta)$.
for $j = 1$ to N do
    Sample $\Theta_l^{(j)} \sim \pi(\phi_l) \times \pi(\theta_l)$ for $l = 1, \dots, L$.
    Compute likelihood $\mathcal{L}\left(\Theta^{(j)}\right)$ using Eq. (5).
end for
NS loop
while not converged do
    Identify the worst point: $\Theta_{\text{worst}}^{(k)}$ with $\mathcal{L}_{\text{worst}}^{(k)}$.
    Remove $\Theta_{\text{worst}}^{(k)}$ from the set of live points.
    Sample a new point $\Theta_{\text{new}} \sim \pi\left(\Theta|\mathcal{L}(\Theta) > \mathcal{L}_{\text{worst}}^{(k)}\right)$.
    Replace $\Theta_{\text{worst}}^{(k)}$ with Θ_{new}.
    Update evidence $\mathcal{Z}$ and weights using discarded $\Theta_{\text{worst}}^{(k)}$.
    Check convergence using $\Delta\mathcal{Z}_k$.
end while
Finalize Evidence:
Add contributions from the final set of live points to $\mathcal{Z}$.
    return Θ estimates, $\mathcal{Z}$.

4. Results and discussions

The practical effectiveness of NS for parameter estimation is most compellingly demonstrated through simulation studies. In this section, we present the results of a case study in which NS is employed to estimate the DoA of signals from two distinct sources using a sensor array.

The sensor array is composed of M identical omnidirectional sensors arranged in a specific geometry, a 3-dimensional Uniform Circular Array (3-D UCA) as depicted in **Figure 2**. This design, proposed by Keskin and Filik [26], is known for its adeptness in differentiating signal directions, a task more challenging for linear or planar sensor arrays.

This case study details the application of NS for DoA estimation using a 3-D UCA comprising eight sensors ($M = 8$). To optimize array performance and prevent grating lobes, the element spacing was set at half the wavelength ($\lambda/2$). The study primarily

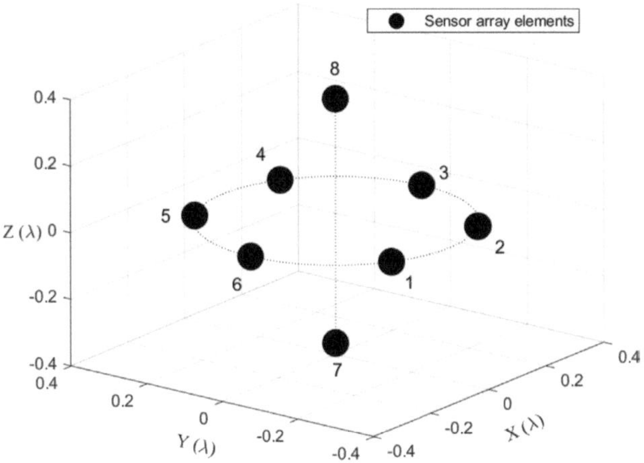

Figure 2.
The spatial positioning of the $M = 8$ sensors in 3D UCA configuration. The axes are labeled in wavelengths to provide a scale reference relative to the signal's wavelength.

focuses on the precise estimation of DoAs for two source signals. These signals, at a carrier frequency of 500 MHz, impinge on the array from directions $\Theta_1 = (30^\circ, 60^\circ)$ and $\Theta_2 = (120^\circ, 25^\circ)$. The simulation process involved 256 snapshots and maintained a signal-to-noise ratio (SNR) of 20 dB to facilitate accurate DoA estimations.

The NS algorithm is initialized with a set of live points drawn from the prior distribution, which, in this case, is uniform over the feasible intervals of azimuth and elevation, given as in (7) and (8), respectively. The simulation iterates over the NS loop, replacing low-likelihood points with new samples (14), until convergence criteria based on the Bayesian evidence Z are met (17).

Figure 3 illustrates a visual representation of our attempt to estimate the DoAs using NS, which successfully estimated the azimuth and elevation of the two signals. The estimated values, represented by the red cross markers on the contour plot, closely match the actual DoAs of the signals, indicated by the green circle markers. This close correspondence between the red and green markers demonstrates the accuracy of NS. In an ideal scenario, the red and green markers would completely overlap, signifying a perfect estimation.

To further assess the accuracy, we analyzed the distribution of estimates using the histograms flanking the contour plot. These histograms, resembling mountain ranges, reveal how frequently estimates fall within specific angular ranges. Essentially, they represent a popularity contest for angles. The left histogram focuses on azimuth angles, while the right histogram deals with elevation angles. In both histograms, we observe peaks indicating the concentrations of estimates, represented by the dashed red lines. The actual signal directions are marked by solid green lines.

Comparing the estimated and actual signal directions, we find that estimates at $\hat{\Theta}_1 = (30.9^\circ, 58.89^\circ)$ and $\hat{\Theta}_2 = (122.9^\circ, 23.847^\circ)$ are quite close to the true directions of $\Theta_1 = (30^\circ, 60^\circ)$ and $\Theta_2 = (120^\circ, 25^\circ)$, respectively. The smaller the gap between the red and green lines in the histograms, the more accurate the estimation method has performed.

Overall, the results demonstrate that NS algorithm effectively estimates the azimuth and elevation of the two signals. The close match between estimated and actual signal directions, along with the narrow peaks in the histograms, confirms the accuracy of the method.

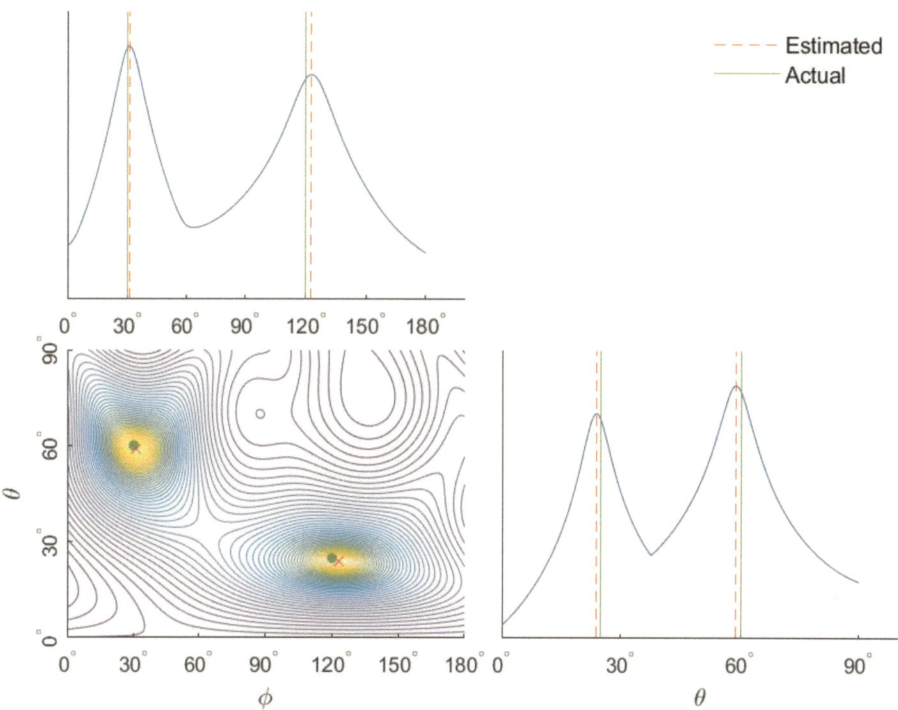

Figure 3.
Parameter estimation using NS for azimuth and elevation. The contour plot shows the posterior distribution of the parameters. The estimated directions of the signals are indicated by red crosses, while green circles depict their actual positions. The histograms show the marginal distributions of the parameters. Dashed red lines represent estimations, whereas solid green lines represent actual positions.

5. Conclusion

The application of NS to DoA estimation represents an important advancement in array signal processing. This case study underscored the robustness of NS in the context of multi-source signals, demonstrating its ability to provide highly accurate DoA estimates under conditions of limited snapshots and elevated noise levels. The iterative nature of the algorithm was a highlight, showing consistent convergence to the actual signal sources with notable accuracy.

The results from this study illuminate the potential of NS as a viable alternative to conventional DoA estimation techniques, especially in environments characterized by complex distributions and high-dimensional parameter spaces. The strong alignment between the estimated and the actual signal directions, as depicted in the detailed contour plots and histograms, validates the algorithm's effectiveness.

Looking ahead, the promise of NS extends to a wider range of signal-processing applications. Its ability to navigate complex posterior distributions and efficiently compute marginal likelihoods suggests its suitability for more sophisticated array configurations and signal-processing tasks. As computing power increases, the intensive computational requirements of NS will become less of a barrier, potentially opening the door to its wider implementation in practical signal-processing scenarios. This investigation lays the groundwork for future research and suggests that, with further refinement and optimization, NS could become an indispensable tool in scientific and engineering fields that rely on signal processing.

Conflict of interest

The authors declare no conflict of interest.

Author details

Fesih Keskin
Department of Computer Engineering, Igdir University, Igdir, Turkiye

*Address all correspondence to: fesih.keskin@igdir.edu.tr

References

[1] Skilling J. Nested sampling for general Bayesian computation. Bayesian Analysis. 2006;**1**(4):833-859. DOI: 10.1214/06-BA127

[2] Henderson RW, Goggans PM, Cao L. Combined-chain nested sampling for efficient Bayesian model comparison. Digital Signal Processing. 2017;**70**:84-93. Available from: https://www.sciencedirect.com/science/article/pii/S1051200417301719

[3] Feroz F, Hobson MP, Bridges M. MultiNest: An efficient and robust Bayesian inference tool for cosmology and particle physics. Monthly Notices of the Royal Astronomical Society. 2009; **398**:1601-1614

[4] Veitch J, Raymond V, Farr B, Farr W, Graff P, Vitale S, et al. Parameter estimation for compact binaries with ground-based gravitational-wave observations using the LALInference software library. Physical Review D. 2015;**91**:042003. DOI: 10.1103/PhysRevD.91.042003

[5] Higson E, Handley W, Hobson M, Lasenby A. Dynamic nested sampling: An improved algorithm for parameter estimation and evidence calculation. Statistics and Computing. 2019;**29**(5): 891-913. DOI: 10.1007/s11222-018-9844-0

[6] Van Trees HL. Optimum array processing: Part IV of detection, estimation, and modulation theory. New York: John Wiley & Sons; 2002

[7] Schmidt R. Multiple emitter location and signal parameter estimation. IEEE Transactions on Antennas and Propagation. 1986;**34**(3): 276-280

[8] Roy R, Kailath T. ESPRIT-estimation of signal parameters via rotational invariance techniques. IEEE Transactions on Acoustics, Speech, and Signal Processing. 1989;**37**(7):984-995

[9] Feroz F, Hobson MP, Cameron E, Pettitt AN. Importance nested sampling and the MultiNest algorithm. The Open Journal of Astrophysics. 2019;**11**:2

[10] Aitken S, Akman OE. Nested sampling for parameter inference in systems biology: Application to an exemplar circadian model. BMC Systems Biology. 2013;7(1):72. DOI: 10.1186/1752-0509-7-72

[11] Speagle JS. Dynesty: A dynamic nested sampling package for estimating Bayesian posteriors and evidences. Monthly Notices of the Royal Astronomical Society. 2020;**493**(3): 3132-3158

[12] Buchner J. Nested sampling methods. Statistics Surveys. 2023;**17**: 169-215. DOI: 10.1214/23-SS144

[13] Tawara N, Ogawa T, Watanabe S, Kobayashi T. Nested Gibbs sampling for mixture-of-mixture model and its application to speaker clustering. APSIPA Transactions on Signal and Information Processing. 2016;**5**:e16

[14] Buchner J. Collaborative nested sampling: Big data versus complex physical models. Publications of the Astronomical Society of the Pacific. 2019;**131**(1004):108005. DOI: 10.1088/1538-3873/aae7fc

[15] Alsing J, Handley W. Nested sampling with any prior you like. Monthly Notices of the Royal Astronomical Society: Letters. 2021; **505**(1):L95-L99

[16] Ashton G, Bernstein N, Buchner J, Chen X, Csányi G, Fowlie A, et al. Nested sampling for physical scientists. Nature Reviews Methods Primers. 2022; **2**(1):39. DOI: 10.1038/s43586-022-00121-x

[17] Vitale S, Del Pozzo W, Li TGF, Van Den Broeck C, Mandel I, Aylott B, et al. Effect of calibration errors on Bayesian parameter estimation for gravitational wave signals from inspiral binary systems in the advanced detectors era. Physical Review D. 2012;**85**:064034. DOI: 10.1103/PhysRevD.85.064034

[18] Allison R, Dunkley J. Comparison of sampling techniques for Bayesian parameter estimation. Monthly Notices of the Royal Astronomical Society. 2013; **437**(4):3918-3928. DOI: 10.1093/mnras/stt2190

[19] Higson E, Handley W, Hobson M, Lasenby A. Sampling errors in nested sampling parameter estimation. Bayesian Analysis. 2018;**13**(3):873-896. DOI: 10.1214/17-BA1075

[20] Thrane E, Talbot C. An introduction to Bayesian inference in gravitational-wave astronomy: Parameter estimation, model selection, and hierarchical models. Publications of the Astronomical Society of Australia. 2019;**36**:e010

[21] Paulen R, Gomoescu L, Chachuat B. Nested sampling approach to set-membership estimation. IFAC-Papers OnLine. 2020;**53**(2):7228-7233 21st IFAC World Congress. Available from: https://www.sciencedirect.com/science/article/pii/S2405896320308545

[22] Jasa T, Xiang N. Nested sampling applied in Bayesian room-acoustics decay analysisa. The Journal of the Acoustical Society of America. 2012; **132**(5):3251-3262. DOI: 10.1121/1.4754550

[23] Fackler CJ, Xiang N, Horoshenkov KV. Bayesian acoustic analysis of multilayer porous media. Journal of the Acoustical Society of America. 2018;**144**(6):3582-3592. Available from: https://eprints.whiterose.ac.uk/141044/, © 2018 Acoustical Society of America. This is an author produced version of a paper subsequently published in Journal of the Acoustical Society of America. Uploaded in accordance with the publisher's self-archiving policy

[24] Landschoot CR, Xiang N. Model-based Bayesian direction of arrival analysis for sound sources using a spherical microphone array. The Journal of the Acoustical Society of America. 2019;**146**(6):4936-4946. DOI: 10.1121/1.5138126

[25] Krim H, Viberg M. Two decades of array signal processing research: The parametric approach. IEEE Signal Processing Magazine. 1996;**13**(4):67-94

[26] Keskin F, Filik T. An optimum volumetric array design approach for both azimuth and elevation isotropic DOA estimation. IEEE Access. 2020;**8**: 183903-183912

Chapter 4

Bayesian Inference for Regularization and Model Complexity Control of Artificial Neural Networks in Classification Problems

Son T. Nguyen, Tu M. Pham, Anh Hoang, Linh V. Trieu and Trung T. Cao

Abstract

Traditional neural network training is usually based on the maximum likelihood to obtain the appropriate network parameters including weights and biases given the training data. However, if the available data are finite and noisy, the maximum likelihood-based network training can cause the neural network after being trained to overfit the noisy data. This problem has been overcome by using the Bayesian inference applied to the neural network training in various applications. The Bayesian inference can allow values of regularization parameters to be found using only the training data. In addition, the Bayesian approach also allows different models (e.g., neural networks with different numbers of hidden units to be compared using only the training data). Neural networks trained with Bayesian inference can be also known as Bayesian neural networks (BNNs). This chapter focuses on BNNs for classification problems with considerations of model complexity of BNNs conveniently handled by a method known as the evidence framework.

Keywords: Bayesian inference, artificial neural networks, classification problems, model complexity, evidence framework

1. Introduction

Multi-layer perceptron (MLP) neural networks are commonly used in various applications of neural networks. The generalization of the neural network after being trained or how well the trained network can have predictions with the new cases is the key challenge when developing the MLP neural network. Indeed, an insufficient network complexity can result in "underfitting" phenomena, in which significant data may be ignored. In contrast, an excessive network complexity can cause "overfitting", in which the noisy data are also fitted. The complexity of the MLP neural network depends on magnitudes of weights and biases. In addition, the network size depending on the number of hidden nodes can be also known as a useful factor to measure the network complexity.

For a long period, the Bayesian approach has been used to improve the generalization capabilities of MLP neural networks despite limited or noisy data [1–3]. Instead of the consideration of magnitudes of weights and biases only, the Bayesian technique considers the distribution of weights and biases which can result a good generalization for the network after being trained. Neural networks trained based on the Bayesian inference are also known as Bayesian neural networks (BNNs).

Until now, BNNs have been utilized in many important applications: fault diagnosis of power transformers using dissolved gas analysis [4], protein secondary structure prediction [5], hands-free control of power wheelchairs for severely disabled people [6–8], detection of hypoglycaemia in children with type 1 diabetes [9, 10], fault identification in cylinders [11], near-infrared spectroscopy [12], and electric load forecasting [13].

As BNNs can be effective for various classification problems, this chapter aims at providing a procedure for deploying BNNs for classification with the following items:

- Exploring the evidence framework to allow all available data to be used to train the BNNs. This work is needed when the collection of relatively large data is expensive or time-consuming.

- As the traditional gradient descent algorithm with a fixed step size and search direction to search local minima usually causes an excessively large network training time, this chapter also focuses on three advanced optimization training algorithms for BNNs including conjugate gradient, scaled conjugate gradient and quasi-Newton algorithms to automatically adjust the step size and search direction to optimal values.

- As finding the optimal network architecture is often a time-consuming task, this chapter also mentions the concept of the Bayesian model comparison to choose the best network architecture. This work is done by evaluating the evidence for different candidature BNNs with varied numbers of hidden nodes. The network architecture which can give the highest value of the evidence will be chosen for the final use.

2. Bayesian learning for neural networks

The objective of maximum likelihood network learning approaches is to minimize a data error function by finding an appropriate single vector of weights and biases. In contrast, the Bayesian technique considers a probability distribution over the weight and bias space, which represents the relative level of belief in various vectors of weights and biases. This work firstly sets a prior distribution. Using Bayes' theorem, the observed data can be transformed from the prior distribution into a posterior distribution that can be used to assess the network's prediction with new values of the input variables.

DOI: http://dx.doi.org/10.5772/intechopen.1002652

2.1 The prior distribution of weights and biases

The weights and biases of a neural network can be split into G groups. Let W_g denote the number of weights and biases in group g and w_g denote the vector of weights and biases in group g. Conveniently, the prior distribution of the vector of weights and biases in group g can be assumed to be a zero-mean Gaussian distribution as:

$$p\left(w_g|\xi_g\right) = \sqrt{\frac{\xi_g}{2\pi}} \exp\left(-\frac{1}{2}\xi_g {w^g}_2\right) \tag{1}$$

where ξ_g is the hyperparameter to constraint the weights and biases in group g. The prior weight and bias distribution in (1) can be defined by assuming that positive and negative weights and biases are equally frequent and have a finite variance.

The prior distribution of all the vectors of weights and biases w is given by:

$$p(w|\psi) = \frac{1}{Z_W(\psi)} \exp\left(-\sum_{g=1}^{G} \xi_g E_{W_g}\right) \tag{2}$$

where $E_{W_g} = \frac{1}{2}\left\|w_g\right\|^2$ is known as the weight function corresponding to group g and ψ is the vector of hyperparameters having the following form:

$$\psi = [\xi_1 \dots \xi_G]^T \tag{3}$$

$Z_W(\psi)$ is a normalization constant which is given by:

$$Z_W(\psi) = \prod_{g=1}^{G} \left(\frac{2\pi}{\xi_g}\right)^{W_g/2} \tag{4}$$

2.2 The posterior distribution of weights and biases

The weights and biases can be adjusted to their most probable (MP) values given the training data D. We can also compute the posterior weight distribution using Bayes' theorem in the form below:

$$p(w|D,\psi) = \frac{p(D|w,\psi)p(w|\psi)}{p(D|\psi)} \tag{5}$$

Eq. (5) can be expressed in words as follows:

$$\textit{Posterior} = \frac{\textit{Likelihood} \times \text{Prior}}{\textit{Evidence}} \tag{6}$$

The posterior distribution of weights and biases is also assumed to be a Gaussian distribution with the following form:

$$p(w|D,\psi) = \frac{\exp(-S(w))}{Z_S(\psi)} \tag{7}$$

where $Z_S(\psi)$ is called the normalization constant and given by:

$$Z_S(\psi) = \int \exp(-S(w))dw \tag{8}$$

In the Bayesian inference, the most probable vector of weights and biases, w_{MP}, must maximizes the posterior distribution of weights and biases. This task is equivalent to searching for the minimization of a cost function.

2.3 The posterior probability of Hyperparameters

Until now, the most probable vector of weights and biases, w_{MP}, given specific values of the hyperparameters, have not yet been computed. Therefore, the hyperparameters must be firstly determined given the model and data. Again, we can express the posterior distribution of the hyperparameters using Bayes' theorem as follows:

$$p(\psi|D) = \frac{p(D|\psi)p(\psi)}{p(D)} \tag{9}$$

In eq. (9), the denominator $p(D)$ does not depend on the hyperparameters. $p(\psi)$ is the prior distribution of the hyperparameters, which is simply assumed to be a uniform distribution. $p(\psi)$ can be later ignored because to infer the values of the hyperparameters, therefore, we only need to search the values of the hyperparameters to maximize $p(D|\psi)$, which is called "the evidence" in (5). The evidence can be exactly calculated by evaluating the following integral:

$$p(D|\psi) = \int p(D|w, \psi)p(w|\psi)dw \tag{10}$$

where $p(D|w, \psi)$ is the probability of the data, traditionally called "the dataset likelihood" and has the form:

$$p(D|w, \psi) = \exp(-E_D) \tag{11}$$

where $E_D = -\sum_{n=1}^{N}\sum_{k=1}^{c} t_k^n \ln(z_k^n)$ is called the "entropy" data error function with z_k^n is the k-th output corresponding to the n-the training pattern and t_k^n is the target corresponding to the k-th output and the n-the training pattern.
Substituting (11) and (2) into (10) results in:

$$p(D|\psi) = \int \exp(-E_D)\frac{1}{Z_W(\psi)} \exp\left(-\sum_{g=1}^{G} \xi_g E_{W_g}\right)dw = \frac{Z_S(\psi)}{Z_W(\psi)} \tag{12}$$

2.4 The Gaussian approximation to the evidence

To compute $Z_S(\psi)$, the cost function $S(w)$ can be approximated around the most probable weight vector w_{MP} by the quadratic form:

$$S(w) = S(w_{MP}) + \frac{1}{2}(w - w_{MP})^T A(w - w_{MP}) \tag{13}$$

where A is the Hessian matrix of the cost function $S(w)$ evaluated at w_{MP} and is given by:

$$A = H + \sum_{g=1}^{G} \xi_g I_g \tag{14}$$

In (14), $H = \nabla\nabla E_D(w_{MP})$, the Hessian matrix of the data error function E_D evaluated at w_{MP}, and $I_g = \nabla\nabla E_{W_g}$ is a diagonal matrix having ones along the diagonal that picks off weights and biases in the group g of the weights and biases. Since the error function $S(w)$ the negative log probability of the weight posterior probability, $Z_S(\psi)$ in (12) is a Gaussian integral that can be approximated to the form:

$$Z_S(\psi) \approx \exp(-S(w_{MP}))(2\pi)^{W/2}(\det A)^{-1/2} \tag{15}$$

By substituting (15) and (4) into (12), we obtain the evidence as follows:

$$\begin{aligned} p(D|\psi) &= \frac{\exp(-S(w_{MP}))(2\pi)^{W/2}(\det A)^{-1/2}}{\prod_{g=1}^{G}\left(\frac{2\pi}{\xi_g}\right)^{W_g/2}} \\ &= \exp(-S(w_{MP}))(\det A)^{-1/2}\left(\prod_{g=1}^{G}\left(\xi_g\right)^{W_g/2}\right) \end{aligned} \tag{16}$$

Taking the logarithm of (16) gives:

$$\ln p(D|\psi) = -S(w_{MP}) - \frac{1}{2}\ln(\det A) + \sum_{g=1}^{G}\frac{W_g}{2}\ln \xi_g \tag{17}$$

Eq. (17) is the basis for the determination of the hyperparameters.

2.5 Determination of the Hyperparameters

The most probable hyperparameters are determined by taking the derivative of (17) regarding ξ_g $(g = 1, \ldots, G)$.

$$\frac{\partial}{\partial \xi_g}\ln p(D|\psi) = -E_{W_g}^{MP} + \frac{W_g}{2\xi_g} - \frac{1}{2}tr\left(A^{-1}I_g\right) \tag{18}$$

where $E_{W_g}^{MP}$ is the data error function evaluated at the most probable vector of weights and biases w_{MP}. By setting (18) to zero, we can obtain the following relationship:

$$2E_{W_g}^{MP}\xi_g = W_g - \xi_g tr\left(A^{-1}I_g\right) = \gamma_g \tag{19}$$

The right-hand side of Eq. (19) is equal to a value γ_g defined as the number of "well-determined" parameters for the weights and biases in group g. Finally, ξ_g is given by:

$$\xi_g = \frac{\gamma_g}{2E_{W_g}^{MP}} \tag{20}$$

3. Advanced training algorithms for BNNs

The BNN training is equivalent to minimizing a cost function $S(w)$, which is a highly non-linear function of the weights and biases. In addition, the matter of minimizing continuous and differentiable functions of multi-dimensional variables has been widely investigated, and various methods can be directly applicable to the problem of neural network training.

The neural network training includes a sequence of iterative steps. At the m-th step, a step size α_m is made in the search direction d_m to update the weights and biases or minimize the cost function over the weight and biases space as follows:

$$w_{m+1} = w_m + \alpha_m d_m \tag{21}$$

In the conventional gradient descent algorithm, the step size α_m is fixed for every step and is usually called the learning rate η. The algorithm also makes the simplest choice for d_m by setting it to $-g_m$ (negative gradient).

$$w_{m+1} = w_m - \eta g_m \tag{22}$$

One of the disadvantages of the gradient descent technique is that it requires the choice of a suitable value for the learning rate η. If setting the value of η "too" large, then the function value may increase. If the setting η sufficiently small, the function value can decrease steadily. However, if very small reductions in the function values are taken, then any limiting sequence may not even be a local minimum. To overcome this drawback, a procedure called "line search" should be used to exactly find a local minimum of the cost function, given the search direction.

One very simple technique for reducing iterative steps is to use a "momentum" term μ $(0 < \mu \leq 1)$ as follows:

$$w_{m+1} = w_m - \eta g_m + \mu(w_m - w_{m-1}) \tag{23}$$

The use of momentum can result in a significant improvement in the performance of gradient descent. However, the inclusion of this term also may give a second parameter needed to be chosen appropriately. Therefore, it is necessary to consider algorithms that can properly choose the step size and search direction. This section presents three advanced optimization training algorithms for BNNs in classification problems. They are conjugate gradient, scaled conjugate gradient, and quasi-Newton algorithms, which belong to a class of automated non-linear parameter optimization techniques. In contrast with the gradient descent algorithm, which depends on parameters specified by the user, these algorithms can automatically adjust the step size and search direction to obtain fast convergence for network training.

3.1 Line search

The line search is required for the conjugate gradient algorithm and quasi-Newton algorithm. The line search is a process for searching the step size $\alpha_{\min}$ to minimize the cost function $S(w)$ given the search direction d_m. A line search consists of two stages:

- Bracketing the minimum: this stage aims to find an interval that includes a triple $a < b < c$ such that $S(a) > S(b)$ and $S(c) > S(b)$. Since the cost function is continuous, this always ensures that there is a local minimum somewhere in the interval (a, b).

- Locating the minimum itself: since the cost function is smooth and continuous, the minimum can simply be obtained by a process of parabolic interpolation. This involves fitting a quadratic polynomial to the cost function through three successive points $a < b < c$ and then moving to the minimum of the parabola.

3.1.1 Bracketing the minimum

For a given bracketing triple $\{a, b, c\}$, we wish to find a new bracket as narrow as possible, as shown in **Figure 1**. This task requires choosing a new trial point $x \in (a, c)$ and verifying this point. The position of the new trial point x can be determined using the "golden section search":

$$\frac{c-x}{c-a} = 1 - \frac{1}{\varphi} \tag{24}$$

where φ is called the "golden ratio".

$$\varphi = \frac{1}{2}\left(1 + \sqrt{5}\right) = 1.6180339887 \tag{25}$$

If $x > b$ then the new bracketing triple is $\{a, b, x\}$ if $S(x) > S(b)$ and is $\{b, x, c\}$ if $S(x) < S(b)$. A similar choice is made if $x < b$. This algorithm is very robust since no assumption is made about the function being minimized, other than continuity.

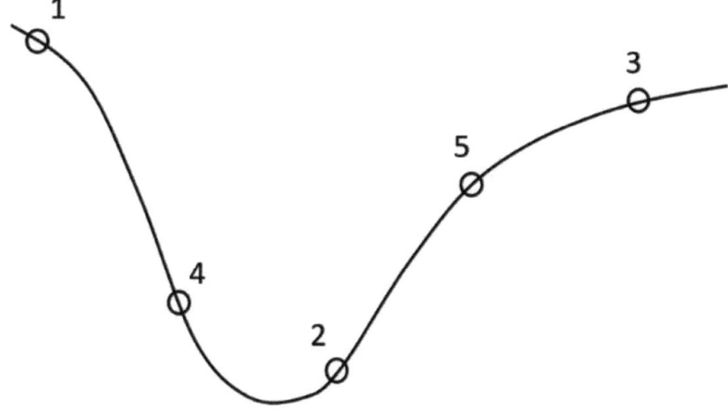

Figure 1.
Golden section search: Initial bracket (1, 2, 3) becomes (4, 2, 3), (4, 2, 5), etc.

3.1.2 Locating the minimum itself

As shown in **Figure 2**, the minimum point of the cost function inside the interval (a, c) can be approximated as a point d that is the minimum of a parabola fitted through three points $(a, S(a))$, $(b, S(b))$ and $(c, S(c))$ as:

$$d = b - \frac{1}{2} \frac{(b-a)^2[S(b) - S(c)] - (b-c)^2[S(b) - S(a)]}{(b-a)[S(b) - S(c)] - (b-c)[S(b) - S(a)]} \tag{26}$$

However, Eq. (26) does not guarantee that d always lies inside the interval (a, c). Also, if b is already at or near the minimum of the interpolating parabola, then the new point d is close to b. This causes slow convergence if b is not close to the local minimum. To avoid this problem, this approach must be combined with the golden section search for finding the exact local minimum.

3.2 Conjugate-gradient training algorithm

In the conjugate-gradient training algorithm, the new search direction is chosen so that [14]:

$$d_m A_m d_{m-1} = 0 \tag{27}$$

d_m is the search direction at step m and d_{m-1} is the search direction at step $m - 1$. A_m is the Hessian matrix of the cost function evaluated at w_m. We say that d_m is *conjugate* to d_{m-1}. If the cost function has the quadratic form, the step size d_m can be written in the form:

$$\alpha_m = \frac{g_m^T d_m}{d_m^T A_m d_m} \tag{28}$$

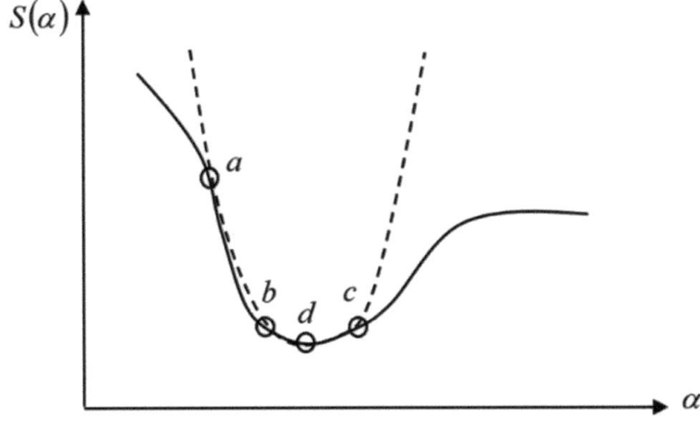

Figure 2.
An illustration of the process of parabolic interpolation used to perform line search minimization: A parabola (shown dotted) is fitted to the three points a, b, c. The minimum of the parabola, at d, gives an approximation to the minimum of S(α).

According to Eq. (28), the determination of α_m requires the evaluation of the Hessian matrix A_m. However, α_m can be found by performing a "very accurate" line search. The new search direction is then given by:

$$d_{m+1} = -g_{m+1} + \beta_m d_m \tag{29}$$

where

$$\beta_m = \frac{\left(g_{m+1} - g_m\right)^T g_{m+1}}{g_m^T g_m} \tag{30}$$

Eq. (30) is called the Polak-Ribiere expression [2, 3].
Finally, the key steps of the algorithm can be summarized as follows:

1. Choosing an initial weight vector w_1.

2. Evaluating the gradient vector g_1 and setting the initial search direction $d_1 = -g_1$.

3. At step m, minimizing $S(w_m + \alpha d_m)$ regarding α to give $w_{m+1} = w_m + \alpha_{\min} d_m$.

4. Testing to see whether the stopping criterion is satisfied.

5. Evaluating the new gradient vector g_{m+1}.

6. Evaluating the new search direction using Eqs. (29) and (30).

3.3 Scaled conjugate-gradient training algorithm

The use of line search allows the step size in the conjugate gradient algorithm to be chosen without having to evaluate the Hessian matrix. However, the line search itself causes some problems. Every line minimization involves several computationally expensive error function evaluations.

The scaled conjugate gradient algorithm avoids the line-search procedure of the conventional conjugate gradient algorithm [15]. To make the denominator of (28) always positive to avoid the increase in the value of the cost function, a multiple of the identity matrix is added to the Hessian matrix in the denominator of (28) to obtain the following parameter:

$$\alpha_m = \frac{g_m^T d_m}{d_m^T A_m d_m + \lambda_m \|d_m\|^2} \tag{31}$$

where λ_m is a "non-negative scale parameter" for adjusting the step size α_m. The denominator in (31) can be written as:

$$\delta_m = d_m^T A_m d_m + \lambda_m \|d_m\|^2 \tag{32}$$

If $\delta_m < 0$, we can increase λ_m to make $\delta_m > 0$. Let the raised value of λ_m be $\overline{\lambda}_m$. Then the corresponding raised value of δ_m is given by:

$$\overline{\delta}_m = \delta_m + \left(\overline{\lambda}_m - \lambda_m\right)\|d_m\|^2 \tag{33}$$

To make $\overline{\delta}_m > 0$, we need to choose:

$$\overline{\lambda}_m = 2\left(\lambda_m - \frac{\delta_m}{\|d_m\|^2}\right) \tag{34}$$

Substituting (34) into (33) gives

$$\overline{\delta}_m = -\delta_m + \lambda_m\|d_m\|^2 \tag{35}$$

$\overline{\delta}_m$ is now positive. This value is used as the denominator in (31) to compute the step size α_m. To find λ_{m+1}, a comparison parameter is firstly defined as:

$$\Delta_m = 2\left[\frac{S(w_m) - S(w_m + \alpha_m d_m)}{\alpha_m d_m^T d_m}\right] \tag{36}$$

Then the value of λ_{m+1} can be adjusted using the following prescriptions:

$$\text{If } \Delta_m > 0.75 \text{ then } \lambda_{m+1} = \frac{\lambda_m}{2} \tag{37}$$

$$\text{If } \Delta_m < 0.25 \text{ then } \lambda_{m+1} = 4\lambda_m \tag{38}$$

In Eq. (31), $d_m^T A d_m$ can be approximated as:

$$d_m^T A_m d_m \approx d_m^T \left[\frac{\nabla S(w_m + \sigma d_m) - \nabla S(w_m)}{\sigma}\right] \tag{39}$$

where $\sigma = \frac{\sigma_0}{\|d_m\|}$ and σ_0 is chosen to be a very small value.
By substituting (39) into (31), we obtain:

$$\alpha_m = \frac{g_m^T d_m}{d_m^T\left[\frac{\nabla S(w_m + \sigma d_m) - \nabla S(w_m)}{\sigma}\right] + \lambda_m\|d_m\|^2} \tag{40}$$

Eq. (40) is used to scale the step size. The new search direction is also determined using the same procedure in the conjugate gradient algorithm.

3.4 Quasi-Newton training algorithm

In the Newton method, the network weights can be updated as:

$$w_{m+1} = w_m - A_m^{-1} g_m \tag{41}$$

The vector $-A_m^{-1} g_m$ is called the "Newton direction" or the "Newton step". However, the evaluation of the Hessian matrix can be very computational. If the cost function is not quadratic, the Hessian matrix may be no longer positive definite causing an increase in the cost function value.

From Eq. (41), we can form the relationship between the weight vectors at steps m and $m+1$ as:

$$w_{m+1} = w_m - \alpha_m F_m g_m \tag{42}$$

From (42), if $\alpha_m = 1$ and $F_m = A_m^{-1}$, we have the Newton method, while if $F_m = I$, we have gradient descent with learning rate α_m.

F_m can be chosen to approximate the Hessian matrix. In addition, F_m must be positive definite so that for small α_m we can obtain a descent method. In practice, the value of α_m can be found by a line search. Eq. (42) is known as the quasi-Newton condition. The most successful method to compute F_m is the Broyden-Fletcher--Goldfarb-Shanno (BFGS) formula [2, 3]:

$$F_{m+1} = F_m + \frac{pp^T}{p^T v} - \frac{(F_m v)v^T F_m}{v^T F_m v} + \left(v^T F_m v\right) u u^T \tag{43}$$

where p, v and u are defined as:

$$p = w_{m+1} - w_m \tag{44}$$

$$v = g_{m+1} - g_m \tag{45}$$

$$u = \frac{p}{p^T v} - \frac{F_m v}{v^T F_m v} \tag{46}$$

Now, the line search with great accuracy is no longer required as the line search does not form a critical factor in the algorithm. For the conjugate gradient algorithm, the line search needs to be performed accurately to ensure that the search direction is set correctly.

4. Bayesian model comparison

When the Bayesian inference is applied to neural network training, the principle of Bayesian model comparison can be utilized to select the optimal number of hidden nodes in the neural network, given the training data. Suppose that there are a set of neural networks X_i with different numbers of hidden nodes. According to Bayes' theorem, we can write down the posterior probabilities of the network X_i, once the training data set D has been observed, in the form:

$$P(X_i|D) = \frac{p(D|X_i)P(X_i)}{p(D)} \tag{47}$$

where $P(X_i)$ is the prior probability of the network X_i and the quantity $p(D|X_i)$ is referred to as the "evidence" for X_i. If there is no reason to assign different priors to the candidature neural networks, the relative probabilities of the networks can be compared based on their evidence.

The evidence of the network X_i can be precisely computed as:

$$p(D|X_i) = \int p(D|w, X_i)p(w|X_i)dw \tag{48}$$

If the posterior distribution is sharply peaked in the weight space around the most probable weight vector, w_{MP}, then the integral (48) can be approximated as:

$$p(D|X_i) \approx p(D|w_{MP}, X_i)\left(\frac{\Delta w_{posterior}}{\Delta w_{prior}}\right) \tag{49}$$

where Δw_{prior} and $\Delta w_{posterior}$ are the prior and posterior uncertainties in the weights. The ratio $\Delta w_{posterior}/\Delta w_{prior}$ is called the Occam factor.

Eq. (49) can be expressed in words as follows:

$$Evidence = Best\ fit\ likelihood \times Occam\ factor \tag{50}$$

The best-fit likelihood measures how well the network fits the data and the Occam factor (<1) penalizes the network for having the weight vector w. A network that has a large best-fit likelihood will receive a large contribution to the evidence. However, if the network has many hidden nodes, then the Occam factor will be very small. Therefore, the network with the largest evidence will make a trade-off between needing a large likelihood to fit the data well and a relatively large Occam factor, so that the model is not too complex.

The logarithm of evidence for the network X_i containing the weight vector w and G hyperparameters ξ_g, at the most probable weight vector, w_{MP}, is given by:

$$\ln Ev(X_i) = -E_D(w_{MP}) + \ln Occ(w_{MP}) + \sum_{g=1}^{G} \ln Occ\left(\xi_g^{MP}\right) \tag{51}$$

where $E_D(w_{MP})$ is the data error evaluated at w_{MP}. $Occ(w_{MP})$ and $Occ\left(\xi_g^{MP}\right)$ are the Occam factors for the weight vector and the hyperparameters at w_{MP}.

4.1 The Occam factor for the weights and biases

The Occam factor for the weights and biases with the Gaussian distributions is given by:

$$Occ(w) = \left[\exp\left(-\sum_{g=1}^{G} \xi_g^{MP} E_{W_g}^{MP}\right)\right]\left(\prod_{g=1}^{G}\left(\xi_g^{MP}\right)^{W_g/2}\right)(\det A)^{-1/2} \tag{52}$$

where ξ_g^{MP} is the most probable value of the hyperparameter for the weights and biases in group g evaluated at w_{MP} and $E_{W_g}^{MP}$ is the weight error in group g evaluated at w_{MP}.

We now consider a two-layer perceptron network with "tanh" activation functions in the hidden layer. There are two kinds of symmetries in the network:

- Firstly, by changing the signs of all incoming and outgoing weights of a hidden node, we can obtain an identical mapping from the input nodes to the output nodes. For a network with M hidden nodes, there are 2^M equivalent weight vectors that result in the same mapping from the inputs to the outputs of the network.

- Secondly, interchanging the values of all incoming and outgoing weights of a hidden node with the corresponding values of the weights associated with another hidden node also results in an identical mapping. For a network with M hidden nodes, there are $M!$ of those permutations.

Therefore, a minimum value of the cost function $S(w)$ can be obtained with $2^M M!$ equivalent weight vectors. Hence, $Occ(w)$ in Eq. (52) should be multiplied by $2^M M!$ for a fully connected network:

$$Occ(w) = \left[\exp\left(-\sum_{g=1}^{G} \xi_g^{MP} E_{W_g}^{MP}\right)\right]\left(\prod_{g=1}^{G}\left(\xi_g^{MP}\right)^{W_g/2}\right)(\det A)^{-1/2}\left(2^M M!\right) \tag{53}$$

4.2 The Occam factor for the Hyperparameters

For the hyperparameter ξ_g, the logarithm of the evidence for (ξ_g, X_i) is assumed to be a quadratic form in $\ln \xi_g$ as:

$$\ln P\left(D|\xi_g, X_i\right) = \ln P\left(D|\xi_g^{MP}, X_i\right) - \frac{1}{2}\frac{\left(\ln \xi_g - \ln \xi_g^{MP}\right)^2}{\sigma_{\ln \xi_g}^2} \tag{54}$$

where $\sigma_{\ln \xi_g}$ is the variance of the distribution for $\ln \xi_g$, and at $\xi_g = \xi_g^{MP}$ it has the form:

$$\left(\sigma_{\ln \xi_g}\right)^{-2} = -\xi_g^2 \frac{\partial^2}{\partial^2 \xi_g} \ln p\left(D|\xi_g, X_i\right) \tag{55}$$

So

$$\begin{aligned} P(D|X_i) &= \int P\left(D|\xi_g, X_i\right) P\left(\ln \xi_g | X_i\right) d\ln \xi_g \\ &= P\left(D|\xi_g^{MP}, X_i\right) \int \exp\left[-\frac{1}{2}\frac{\left(\ln \xi_g - \ln \xi_g^{MP}\right)^2}{\sigma_{\ln \xi_g}^2}\right] \frac{1}{\ln \Omega} d\ln \xi_g \\ &= P\left(D|\xi_g^{MP}, X_i\right) Occ\left(\xi_g\right) \end{aligned} \tag{56}$$

where

$$Occ\left(\xi_g\right) = \frac{\sqrt{2\pi}\sigma_{\ln \xi_g}}{\ln \Omega} \tag{57}$$

The parameter Ω can be set to a specific value, for example 10^3, which indicates a subjective estimate of the hyperparameter. However, in practice, Ω is only a minor factor, as it is the same for every network. Finally, $Occ\left(\xi_g\right)$ denotes the Occam factor for the hyperparameter ξ_g.

In Eq. (57), $\sigma_{\ln \xi_g}$ can be approximated as:

$$\sigma_{\ln \xi_g} \approx \sqrt{\frac{2}{\gamma_g}} \tag{58}$$

Substituting (58) into (57) gives:

$$Occ\left(\xi_g\right) = \frac{\sqrt{4\pi/\gamma_g}}{\ln \Omega} \tag{59}$$

4.3 Combining the terms of evidence

We can now combine the terms of evidence. Firstly, taking the logarithm of (52) gives:

$$\ln Occ(w) = -\sum_{g=1}^{G} \xi_g^{MP} E_{W_g}^{MP} + \sum_{g=1}^{G} \frac{W_g}{2} \ln \xi_g - \frac{1}{2} \ln(\det A) + \ln M! + M \ln 2 \tag{60}$$

Similarly, taking the logarithm of (59) gives:

$$\ln Occ\left(\xi_g^{MP}\right) = \frac{1}{2} \ln\left(\frac{4\pi}{\gamma_g^{MP}}\right) - \ln(\ln \Omega) \tag{61}$$

Substituting (60) and (61) in (51) gives:

$$\begin{aligned} \ln Ev(X_i) = {} & -S(w_{MP}) + \sum_{g=1}^{G} \frac{W_g}{2} \ln \xi_g - \frac{1}{2} \ln(\det A) + \ln M! + M \ln 2 \\ & + \sum_{g=1}^{G} \left[\frac{1}{2} \ln\left(\frac{4\pi}{\gamma_g^{MP}}\right) - G \ln(\ln \Omega)\right] \end{aligned} \tag{62}$$

where W_g is the number of weights and biases in group g. Ω is a factor because it is the same for all models and does not affect the relative comparison of log evidence of different neural networks. Therefore, Eq. (62) can be conveniently used to rank the complexities of different networks [12] and we may expect that the neural network with the highest evidence will provide the best results on unseen data.

5. A case study on power transformer fault classification using dissolved gas analysis

In power generation, transmission, and distribution networks, power transformers are a common type of electrical equipment. Typically, arcing, corona discharges, sparking, and overheating in insulating materials are the results of power transformer defects that are just beginning. In response to these stressors, insulating materials may deteriorate or break down, releasing several gases. Since these dissolved gases can be analyzed, it is possible to learn useful information about the materials and fault conditions that are involved. Power transformer insulating oil dissolved gas analysis (DGA) is a widely used method for assessing the health of power transformers. By analyzing various gas concentration ratios such as Doernenburg ratios, Rogers ratios, and Duval's triangle method, one can perform conventional analysis procedures of dissolved gases.

5.1 Conventional methods of DGA for power transformer insulating oil

The main reasons for gas generation inside a power transformer in operation are evaporation, electrochemical and thermal degradation, and decomposition. Bonds between carbon and hydrogen and carbon and carbon are broken in fundamental chemical processes. The gases hydrogen (H_2), methane (CH_4), acetylene (C_2H_2), ethylene (C_2H_4), and ethane (C_2H_6) can all be produced by combining active hydrogen atoms and hydrocarbon fragments produced by this event. With cellulose insulation, methane (CH_4), hydrogen (H_2), monoxide (CO), and carbon dioxide (CO_2) can be produced via heat decomposition or electrical faults. "Key gases" is the common term for these gases.

Measuring the concentration level (in ppm) of each important gas comes first in the analysis of DGA results after samples of transformer insulating oil are taken. Once critical gas concentrations exceed the usual range, analysis procedures should be employed to identify any potential transformer defects. These methods entail computing important gas ratios and comparing those ratios to recommended limits. The methods based on the following gas ratios—CH_4/H_2, C_2H_2/C_2H_4, C_2H_2/CH_4, C_2H_6/C_2H_2, and C_2H_4/C_2H_6—are Doernenburg ratios and Rogers ratios, respectively. **Tables 1** and **2**, respectively, display the suggested upper and lower bounds for the Doernenburg ratios approach and the Rogers ratios method.

In Duval's triangle approach, the amounts of three important gases—methane (CH_4), acetylene (C_2H_2), and ethylene (C_2H_4)—are calculated. The percentage associated with each gas is then calculated by dividing its concentration by the sum of the amounts of the other three gases. To determine a diagnosis, these results are then

Suggested fault diagnosis	$R_1 = \frac{CH_4}{H_2}$	$R_2 = \frac{C_2H_2}{C_2H_4}$	$R_3 = \frac{C_2H_2}{CH_4}$	$R_4 = \frac{C_2H_6}{C_2H_2}$
Thermal decomposition	>1.0	<0.75	<0.3	>0.4
Partial discharge	<0.1	—	<0.3	>0.4
Arcing	$>0.1-<1.0$	>0.75	>0.3	<0.4

Table 1.
Suggested limits of Doernenburg ratios method.

Suggested fault diagnosis	$R_1 = \frac{CH_4}{H_2}$	$R_2 = \frac{C_2H_2}{C_2H_4}$	$R_5 = \frac{C_2H_4}{C_2H_6}$
Unit normal	$>0.1-<1.0$	<0.1	<1.0
Low-energy density arcing-PD	<0.1	<0.1	<1.0
Arcing-high energy discharge	$0.1-1.0$	$0.1-3.0$	>3.0
Low temperature thermal	$>0.1-<1.0$	<0.1	$1.0-3.0$
Thermal$<700°C$	>1.0	<0.1	$1.0-3.0$
Thermal$>700°C$	>1.0	<0.1	>3.0

Table 2.
Suggested limits of Rogers ratios method.

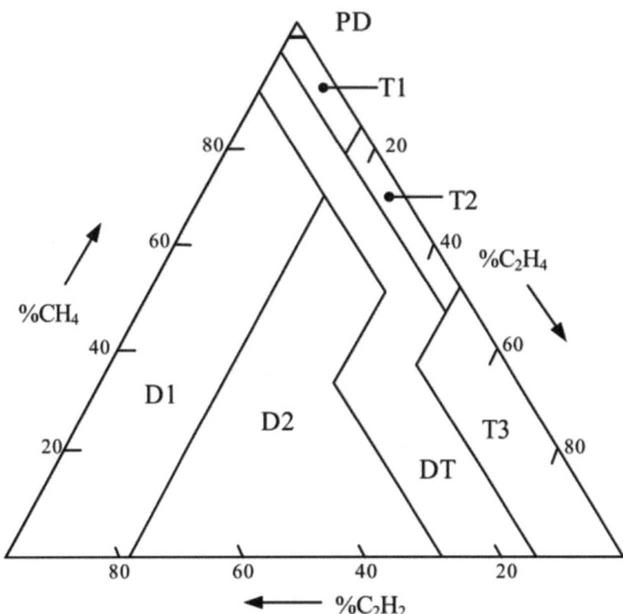

Figure 3.
Duval's triangle.

plotted in Duval's triangle as seen in **Figure 3**. Partially discharged (PD), low-energy discharge (D1), high-energy discharge (D2), thermal fault (T1), thermal fault (T2), and thermal fault (T3) are all indicated by sections of the triangle.

5.2 Results and discussion

BNNs were tested and trained using the IEC TC10 databases. There is an output pattern that corresponds to each input pattern and describes the fault type for a specific diagnosis criterion. In this study, five important gases—hydrogen (H_2), methane (CH_4), ethylene (C_2H_4), ethane (C_2H_6), and acetylene (C_2H_2)—that are all combustible are used. Five fault kinds are indicated by the output vector's codes of 0 and 1, which are shown in **Table 3**. As indicated in **Table 4**, the training set was created by taking 81 data samples, and the test set was created by taking 36 data samples.

Fault type	Output vector
PD	$[1\ 0\ 0\ 0\ 0]^T$
D1	$[0\ 1\ 0\ 0\ 0]^T$
D2	$[0\ 0\ 1\ 0\ 0]^T$
T1 & T2	$[0\ 0\ 0\ 1\ 0]^T$
T3	$[0\ 0\ 0\ 0\ 1]^T$

Table 3.
Fault types and corresponding output vectors.

	Numbers of data samples	
Fault type	**Training set**	**Test set**
PD	5	4
D1	18	8
D2	36	12
T1 & T2	10	6
T3	12	6
Total	81	36

Table 4.
Datasets from the IEC TC 10 database.

Low dissolved gas concentrations of a few ppm (parts per million) are typical for power transformers. But hundreds or tens of thousands of ppm can frequently be brought on by malfunctioning power transformers. The dissolved gas measurements are typically difficult to visualize because of this issue. The order of magnitude of DGA concentrations, rather than their absolute values, can be used to determine the characteristics of DGA data that are the most meaningful. The $\log_{10}$ is a useful way for simple comprehension of DGA data.

To use with the neural network training, the training data needs to be normalized for a range of 0 and 1 by using the following formula:

$$y_i = \frac{x_i - \min(X)}{\max(X) - \min(X)} \tag{63}$$

5.2.1 The network training procedure

Different BNNs with varied numbers of hidden nodes were trained to find the optimal number of hidden nodes (number of nodes in the hidden layer). The following characteristics apply to these networks:

1. The magnitudes of the weights on the connections from the input nodes to the hidden nodes, the biases of the hidden nodes, the weights on the connections from the hidden nodes to the output nodes, and the biased of the output nodes were handled by four hyperparameters ξ_1, ξ_2, ξ_3, and ξ_4.

2. The number of network inputs is determined by the number of gas ratios used in a particular diagnosis procedure, plus one augmented input with a fixed value of 1.

3. As indicated in **Table 3**, there are five outputs, each of which corresponds to a certain category of errors. For a specific number of hidden nodes, 10 neural networks with different initial weights and biases were trained.

The training procedure was implemented with the following steps:

1. The weights and biases in four groups were initialized by random selections from zero-mean and unit variance Gaussians. The hyperparameters were also initialized to be small values.

2. The scaled conjugate gradient technique was used to minimize the cost function.

3. The values of the hyperparameters were re-estimated in accordance with Eqs. (19) and (20) once the cost function has reached a local minimum.

4. Steps 2 and 3 were repeated until either the cost function value is smaller than a predefined value, or the required number of training iterations had reached.

5.2.2 Power transformer fault classification

Power transformer faults can be categorized according to gas ratios including Doernenburg and Rogers ratios.

5.2.2.1 Doernenburg ratios

The input vector of the network in this case is a vector consisting of four elements as follows:

$$[x] = \left[\frac{CH_4}{H_2}, \frac{C_2H_2}{C_2H_4}, \frac{C_2H_2}{CH_4}, \frac{C_2H_6}{C_2H_2}\right]^T \tag{64}$$

The training set was used to train several classification BNNs with various numbers of hidden nodes. Ten BNNs with various randomly chosen beginning weights and biases were trained for a specific number of hidden nodes, and the log evidence was then computed. The networks with two hidden nodes had the maximum log evidence, as seen in **Figure 4**. In addition, **Figure 5** displays the best overall accuracy of fault classification, which is like the highest log evidence in **Figure 4**.

Table 5 illustrates the number of well-determined parameters and the change in four hyper-parameters. According to **Table 6**, the confusion matrix of the optimized BNN that was used to categorize the unknown input vectors can be obtained at a rate of 83.33%.

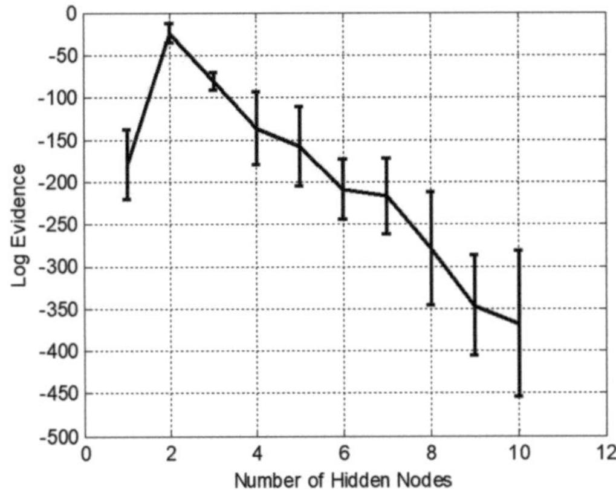

Figure 4.
Log evidence versus the number of hidden nodes (Doernenburg ratios).

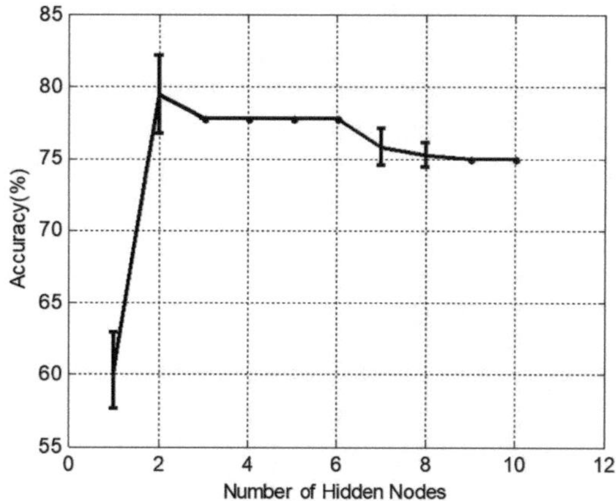

Figure 5.
Overall accuracy in relation to the number of hidden nodes (Doernenburg ratios).

Period	ξ_1	ξ_2	ξ_3	ξ_4	γ
1	0.022	0.044	0.008	0.409	18.555
2	0.039	0.083	0.006	0.753	15.803
3	0.061	0.134	0.005	0.865	15.451

Table 5.
Changes in four hyper-parameters and the number of well-determined parameters over various hyper-parameter re-estimation times (Doernenburg ratios).

		Predicted classification				
Actual classification	Fault	PD	D1	D2	T1&T2	T3
	PD	2	0	0	2	0
	D1	0	5	3	0	0
	D2	0	0	12	0	0
	T1&T2	0	0	0	5	1
	T3	0	0	0	0	6
Accuracy (%)			83.33			

Table 6.
The BNN's confusion matrix for categorizing unknown input vectors (Doernenburg ratios).

5.2.2.2 Rogers ratios

The input network vector in this case is formed based on four gas ratios as follows:

$$[x] = \left[\frac{CH_4}{H_2}, \frac{C_2H_2}{C_2H_4}, \frac{C_2H_4}{C_2H_6}, \frac{C_2H_6}{CH_4} \right]^T \tag{65}$$

The training set was used to train different BNN classifiers with various numbers of hidden nodes. Ten networks were trained with various randomly starting weights and biases for a certain number of hidden nodes, and the log evidence was evaluated. The networks with two hidden nodes can produce the highest log evidence, as seen in **Figure 6**. According to **Figure 7**, this network architecture can also provide the highest overall accuracy of fault classification.

In **Table 7**, the number of well-determined parameters and the change in four hyper-parameters are shown. The optimized BNN's confusion matrix is shown in

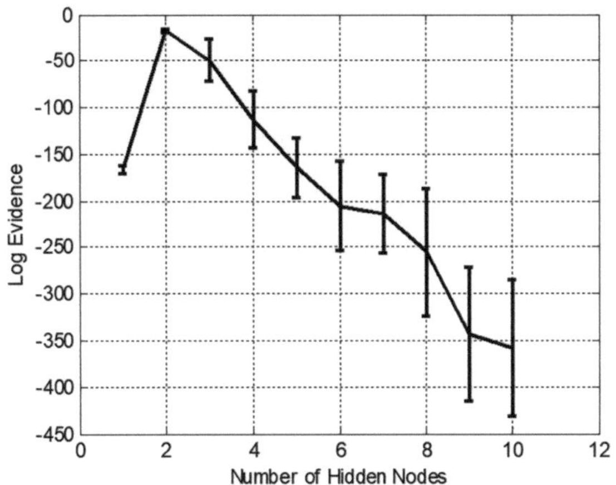

Figure 6.
Log evidence versus the number of hidden nodes (Rogers ratios).

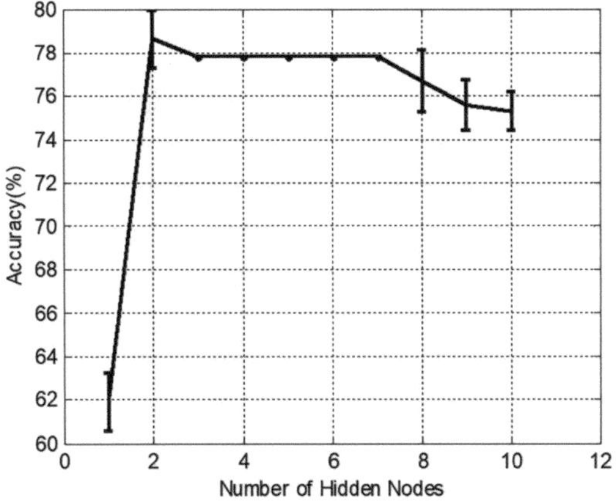

Figure 7.
Overall accuracy in relation to the number of hidden nodes (Rogers ratios).

Period	ξ_1	ξ_2	ξ_3	ξ_4	γ
1	0.026	0.012	0.009	0.268	18.645
2	0.039	0.015	0.007	0.353	16.315
3	0.053	0.02	0.005	0.333	15.801

Table 7.
Four hyper-parameter variations and the number of well-determined parameters based on hyper-parameter re-estimation times (Rogers ratios).

		Predicted classification				
True classification	Fault	PD	D1	D2	T1&T2	T3
	PD	2	0	0	2	0
	D1	0	4	4	0	0
	D2	0	0	12	0	0
	T1&T2	0	0	0	5	1
	T3	0	0	0	0	6
Accuracy (%)			80.56			

Table 8.
Confusion matrix of the trained BNN for identifying unknown input vectors (Rogers ratios).

Doernenburg ratio approach corresponding to suggested gas ratio limits	79.48 (%)
Doernenburg ratio method corresponding to the BNN	83.33 (%)
Rogers ratio method corresponding to suggested gas ratio limits	40.17 (%)
Rogers ratio method corresponding to the BNN	80.56 (%)

Table 9.
Overall accuracy comparison of the suggested gas ratio limit and BNN-based classification techniques.

Table 8, and its overall accuracy in classifying faults from unknown input vectors is 80.56%. **Table 9** compares recommended limit and BNN-based DGA techniques using the same training data set. The recommended limit-based solutions are obviously much outclassed by the BNN-based methods.

The regulariztion parameters (hyperparameters) and the appropriate number of hidden nodes in the network can be easily determined based on the investigation of the Bayesian inference framework for MLP neural network training. The suitability of a BNN design based on a few nodes in the hidden layer for early fault detection in power transformers is demonstrated. The diagnosis criterion under consideration mostly determines how many hidden units are present. For diagnosing power transformer faults, this study also compares proposed gas ratio limit-based approaches with BNN-based methods. This study's future work will involve comparing BNNs and other machine-learning classifiers for power transformer DGA.

6. Conclusions

As standard neural networks that have been expanded with the posterior inference, BNNs can be used to avoid over-fitting. The Bayesian inference can thus be used to determine a probability distribution over a hypothetical neural network. To determine the optimum network architecture and ensure high generalization, the normal neural network training approach can be modified appropriately to directly handle the model comparison problem. Finally, we can confirm that the Bayesian technique can be utilized to avoid overfitting, allow network learning with limited and noisy datasets, and notify us of the level of uncertainty in various predictions.

Acknowledgements

The authors would like to extend their heartfelt appreciation to Professor Ian Nabney (University of Bristol, UK) for his assistance throughout the discovery of the open-source Netlab software used for this study.

Conflict of interest

The authors declare no conflict of interest.

Author details

Son T. Nguyen*, Tu M. Pham, Anh Hoang, Linh V. Trieu and Trung T. Cao
Hanoi University of Science and Technology, Hanoi, Vietnam

*Address all correspondence to: son.nguyenthanh@hust.edu.vn

References

[1] MacKay JC. The evidence framework applied to classification networks. Neural Computation. 1992;**4**:720-736

[2] Bishop CM. Neural Networks for Pattern Recognition. New York: Oxford University Press; 1995

[3] Nabney IT. Netlab: Algorithms for Pattern Recognition (Advances in Pattern Recognition). Springer Nature Customer Service Center LLC; 2002

[4] Thanh SN, Goetz S. Dissolved gas analysis of insulating oil for power transformer fault diagnosis with Bayesian neural network. JST: Smart Systems and Devices. 2022;**32**(3): 61-68

[5] Thanh SN, Johnson CG. Protein secondary structure prediction using an optimized Bayesian classification neural network. In: The 5th International Conference on Neural Computing Theory and Application, Vilamoura, Algarve, Portugal. 20–22 Sep 2013

[6] Thanh SN, Nguyen HT, Taylor PB, Middleton J. Improved head direction command classification using an optimized Bayesian neural network. In: Proceedings of the 28th Annual International Conference of the IEEE Engineering in Medicine and Biology Society, New York City, New York, USA. 31 Aug–3 Sep 2006. pp. 5679-5682

[7] Thanh SN, Tan HN, Taylor P. Bayesian neural network classification of head movement direction using various advanced optimisation training algorithms. In: Proceedings of the First IEEE/RAS - EMBS International Conference on Biomedical Robotics and Biomechatronics, Pisa, Italy. 20–22 Feb 2006. pp. 1-6

[8] Thanh SN, Tan HN, Taylor P. Hands-free control of power wheelchairs using Bayesian neural networks. In: Proceedings of the IEEE Conference on Cybernetics and Intelligent Systems, Singapore. 1–3 Dec 2004. pp. 745-759

[9] Nguyen HT, Ghevondian N, Thanh SN, Jones TW. Detection of hypoglycaemic episodes in children with type 1 diabetes using an optimal Bayesian neural network algorithm. In: Proceedings of the 29th Annual International Conference of the IEEE Engineering in Medicine and Biology Society, Lyon, France. 23–26 Aug 2007

[10] Nguyen HT, Ghevondian N, Thanh SN, Jones TW. Optimal Bayesian neural-network detection of hypoglycaemia in children with type 1 diabetes using a non-invasive and continuous monitor (HypoMon). In: The American Diabetes Association's 67th Scientific Session, Chicago, Illinois, USA. 22–26 Jun 2007

[11] Marwala T. Scaled conjugate gradient and Bayesian training of neural networks for fault identification in cylinders. Computers & Structures. 2001;**79**(32):2793-2803

[12] Thodberg HH. A review of Bayesian neural networks with an application to near infrared spectroscopy. IEEE Transactions on Neural Networks. 1996; **7**(1):56-72. DOI: 10.1109/72.478392

[13] Tito EH, Zaverucha G, Vellasco M, Pacheco M. Bayesian neural networks for electric load forecasting. In: Proceedings. The 6th International Conference on Neural Information Processing, Denver, Colorado, USA. 29 Nov–4 Dec 1999. pp. 407-411. DOI: 10.1109/ICONIP.1999.844023

[14] Charalambous C. Conjugate gradient algorithm for efficient training of artificial neural networks. In: IEE Proceedings G (Circuits, Devices and Systems). Vol. 139, No. 3. The Institution of Electrical Engineers; 1992. pp. 301-310

[15] Møller MF. A scaled conjugate gradient algorithm for fast supervised learning. Neural Networks. 1993;**6**(4): 525-533

Chapter 5

Performance Comparison between Naive Bayes and Machine Learning Algorithms for News Classification

Merve Veziroğlu, Erkan Veziroğlu and İhsan Ömür Bucak

Abstract

The surge in digital content has fueled the need for automated text classification methods, particularly in news categorization using natural language processing (NLP). This work introduces a Python-based news classification system, focusing on Naive Bayes algorithms for sorting news headlines into predefined categories. Naive Bayes is favored for its simplicity and effectiveness in text classification. Our objective includes exploring the creation of a news classification system and evaluating various Naive Bayes algorithms. The dataset comprises BBC News headlines spanning technology, business, sports, entertainment, and politics. Analyzing category distribution and headline length provided dataset insights. Data preprocessing involved text cleaning, stop word removal, and feature extraction with Count Vectorization to convert text into machine-readable numerical data. Four Naive Bayes variants were evaluated: Gaussian, Multinomial, Complement, and Bernoulli. Performance metrics such as accuracy, precision, recall, and F1 score were employed, and Naive Bayes algorithms were compared to other classifiers like Logistic Regression, Random Forest, Linear Support Vector Classification (SVC), Multi-Layer Perceptron (MLP) Classifier, Decision Trees, and K-Nearest Neighbors. The MLP Classifier achieved the highest accuracy, underscoring its effectiveness, while Multinomial and Complement Naive Bayes proved robust in news classification. Effective data preprocessing played a pivotal role in accurate categorization. This work contributes insights into Naive Bayes algorithm performance in news classification, benefiting NLP and news categorization systems.

Keywords: Naive Bayes, machine learning, news classification, natural language processing, data preprocessing

1. Introduction

The exponential growth of digital content in recent years has led to an enormous amount of information available online. As a result, there is a growing need for efficient and effective methods to automatically classify and categorize textual data. News classification, a fundamental application of natural language processing (NLP), aims to automatically assign news headlines to certain predefined categories based on their content. This task has important practical implications such as content recommendation, information retrieval, and sentiment analysis.

In this research, we present a comprehensive News Classification task implemented using the Python programming language. The focus of this work is on leveraging the power of Naive Bayes algorithms to accurately categorize news headlines into different classes. Naive Bayes, a well-established and widely used machine learning algorithm known for its simplicity and effectiveness in text classification tasks, is chosen as the basis of our research.

The main objective of this work is twofold: first, to explore the intricacies of building a robust News Classification system, and second, to rigorously evaluate the performance of various Naive Bayes algorithms on the task at hand. To achieve this, we use a dataset obtained from the reputable BBC News Corpus [1], which contains news headlines from five different categories: technology, business, sports, entertainment, and politics. Through detailed analysis, we aim to gain valuable insights into the distribution of categories and the characteristics of headlines. We then perform extensive data preprocessing, including data cleaning, stop word removal, and feature extraction using the Count Vectorizer technique [2]. The dataset is split into training and test sets, and the transformed numeric vectors serve as input for training Naive Bayes algorithms [3].

Our research examines four variants of Naive Bayes: Gaussian Naive Bayes, Multinomial Naive Bayes, Complement Naive Bayes, and Bernoulli Naive Bayes [4]. Each algorithm undergoes rigorous training and is thoroughly evaluated on test data using well-known performance metrics such as accuracy, precision, recall, and F1 score [5]. We also extensively compare the performance of Naive Bayes algorithms with other popular machine learning classifiers such as Logistic Regression, Random Forest, Linear SVC, MLP Classifier, Decision Tree, and K-Nearest Neighbors (KNN) [6]. A detailed analysis of the results is performed to distinguish the relative strengths and weaknesses of each algorithm in the context of news classification.

The findings of our work show that all Naive Bayes algorithms exhibit commendable accuracy on news classification tasks, with Multinomial Naive Bayes and Complement Naive Bayes standing out as particularly effective variants. Furthermore, the MLP Classifier emerges as the best-performing machine learning classifier, demonstrating its effectiveness in the field of news categorization.

In conclusion, this academic research highlights the competent application of Naive Bayes algorithms in news classification and underlines the importance of rigorous data preprocessing and careful algorithm selection. The comprehensive results and insights from this research can serve as a valuable resource for researchers and practitioners in the field of natural language processing and contribute to the advancement of efficient news categorization systems.

2. Related works

News classification is an important area of research to explore various approaches and algorithms to effectively categorize news articles. Many studies in this area have examined the effectiveness of various machine learning and deep learning algorithms, such as Naive Bayes, Support Vector Machines (SVM), Random Forest, Logistic Regression, and neural networks.

The Naive Bayes algorithm is widely used in news classification because of its simplicity, efficiency, and effectiveness. In recent years, several works have been conducted to investigate the use of Naive Bayes algorithms in news classification and to improve its performance. In this section, we will review some of the recent works that use Naive Bayes algorithms in news classification.

Rana et al. investigated news classification based on news headlines and discussed the effectiveness of the Naive Bayes algorithm in this task. They analyze various approaches to feature selection, preprocessing, and classification and compare the performance of Naive Bayes with other classification algorithms. The authors concluded that the Naive Bayes algorithm is a simple and effective method for news classification based on news headlines [7].

Shahi and Pant applied Naive Bayes, Support Vector Machines (SVM), and Neural Networks (NN) to classify Nepalese news articles. They used TF-IDF vectorization to convert text data into machine-readable format and compared the performance of the three algorithms. The results show that Naive Bayes outperforms SVM and NN in terms of accuracy and F1-score. The authors suggested that Naive Bayes algorithm is a suitable method for Nepalese news classification [8].

Chy et al. used a Naive Bayes classifier to classify Bangla news articles. They applied a stemmer and stop word removal to preprocess the text data and used TF-IDF vectorization to convert the data into machine-readable format. The authors compared the performance of Naive Bayes with other classification algorithms and found that Naive Bayes achieved the highest accuracy. They suggested that the use of a stemmer and stop word removal can improve classification accuracy [9].

Bracewell et al. applied the Naive Bayes algorithm to classify Japanese and English news articles. They used a combination of category classification and topic discovery to classify the articles and compared the performance of Naive Bayes with other classification algorithms. The results showed that Naive Bayes achieved the highest accuracy and F1 score. The authors suggested that Naive Bayes algorithm is a suitable method for cross-lingual news classification [10].

Albahr and Albahar used a publicly available dataset to evaluate the news detection capabilities of various machine learning classifiers. They used 80% of the data in the training process and the remaining 20% in the testing process. Naive Bayes classification achieved the best results among the others [11].

Sristy and Somayajulu proposed a semi-supervised approach using the Naive Bayes algorithm to classify news articles based on their content. They used a small labeled dataset and a large unlabeled dataset to train the classifier and compared the performance of Naive Bayes with other semi-supervised algorithms. The results showed that Naive Bayes achieved the highest accuracy and F1-score. The authors suggested that semi-supervised approaches can improve the accuracy of news classification with limited labeled data [12].

Granik and Mesyurar present a simple approach for news detection using a naive Bayes classifier. This approach was implemented as a software system and tested on a dataset of Facebook news posts. In the tests, approximately 74% classification accuracy was achieved [13].

Overall, these works demonstrate the versatility and effectiveness of Naive Bayes algorithms in news classification. They provide insights into how classification accuracy can be improved using different approaches for feature selection, preprocessing, and classification. They also suggest that using Naive Bayes with other algorithms or techniques can further improve the accuracy of news classification.

3. Methodology

In this section, we outline the research methodology used to conduct the News Classification work and evaluate the performance of Naive Bayes algorithms. The work includes a systematic approach to data collection, data preprocessing, model training, and performance evaluation. In the following, we describe the key steps taken to ensure the rigor and reliability of the research. The methodology steps are shown in **Figure 1.**

3.1 Data collection

The data for this work are collected from the BBC News Corpus [1] dataset, which contains news headlines from various categories. The dataset contains two main columns: category and text. The category column represents a category to which each news item belongs, whereas the text column contains the actual text of the news item.

The dataset consists of news headlines from five different categories: technology, business, sports, entertainment, and politics. The number of data points available in each category is shown in **Table 1.**

This dataset presents a variety of news headlines, allowing for a comprehensive analysis of Naive Bayes algorithms for news classification. The data collection process ensures that the dataset represents a balanced distribution of news headlines in different categories, making it possible to derive reliable and meaningful insights from the work. **Figure 2** shows the details of the dataset.

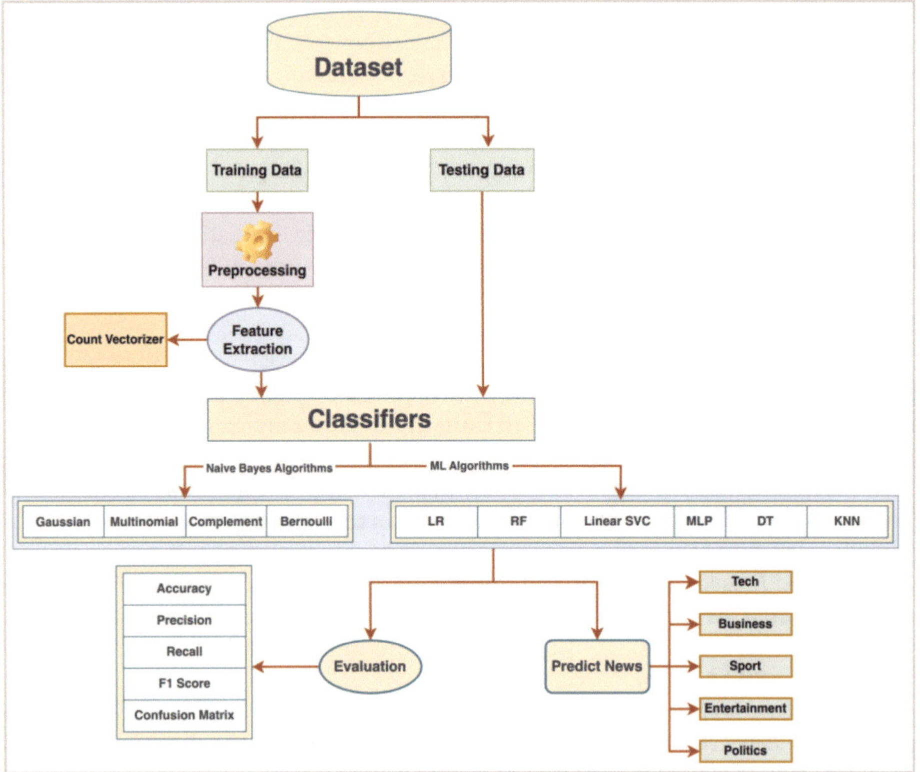

Figure 1.
General operating procedure.

Category	Number of data
Tech	401
Business	510
Sport	511
Entertainment	386
Politics	417
Total	2225

Table 1.
Distribution of the data set.

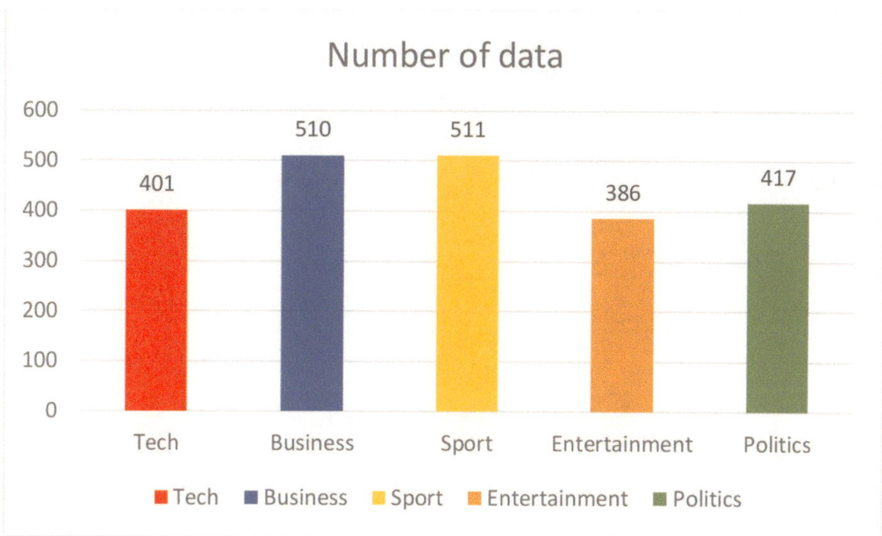

Figure 2.
Data count distributions for each class.

3.2 Data preprocessing

Data preprocessing is a critical step in preparing raw text data for analysis and classification tasks in natural language processing (NLP). In this task, the data preprocessing steps shown in **Figure 3** were performed on the news headlines dataset obtained from the BBC News Corpus. The main goal of data preprocessing is to transform the raw text data into a structured format suitable for further analysis using machine learning algorithms.

3.2.1 Data cleaning

In the data cleaning phase, we first addressed the issue of punctuation and special characters in news headlines. These non-textual elements do not contribute to the overall meaning and context of the text, and can create noise during analysis. Using regular expressions, we effectively removed punctuation and special characters,

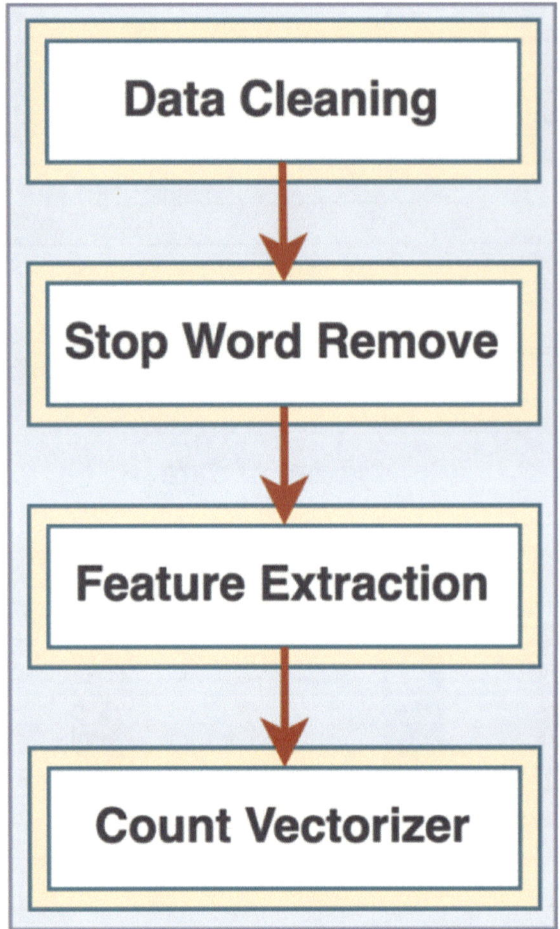

Figure 3.
Model preprocessing techniques.

leaving only the core textual content intact. Some news headlines may contain numeric values, such as dates, statistics, or other numeric data, which are not appropriate for classification purposes. To prevent numeric values from affecting classification models, we removed them from the headlines. This step ensured that the focus remained on textual content only and minimized the impact of irrelevant information.

3.2.2 Stop word remove

Stop words are commonly used words in a language, such as articles ("the," "a," and "an") and other frequent words ("is," "are," and "in"). These words do not carry any significant meaning and are usually removed from the text to reduce dimensionality and focus on more informative features. We used a predefined list of stop words specific to the English language for this purpose. Using the selected stop words, we proceeded to extract them from news headlines. This step effectively reduced the number of words in the dataset, simplifying the subsequent analysis and increasing the relevance of the retained words.

3.2.3 Feature extraction

Tokenization is the process of breaking text into individual words or tokens. We performed tokenization to enable further analysis at the word level by breaking news headlines into meaningful units. From the tokenized words, we created a vocabulary of unique words found in news headlines. This lexicon served as a reference for feature extraction and ensured a consistent representation of words across the dataset. For each news headline, we calculated the term frequency that represents the number of occurrences of each word in that headline. This frequency information was required to convert the textual data into numeric vectors, which were then fed into the machine learning algorithms. The term frequency information was used to create a custom matrix wherein each row represents a news headline, and each column corresponds to a unique word in the vocabulary. The values within the matrix indicate the frequency of occurrence of each word in the relevant headline.

3.2.4 Count vectorizer

The Count Vectorizer technique is a basic approach for creating a numeric matrix representation of text data. It creates a vocabulary from tokenized words and assigns a unique integer to each word. Using the vocabulary generated by Count Vectorizer, a document term matrix was created wherein each row represents a news headline, and each column corresponds to a unique word in the vocabulary. The values in the matrix indicate the frequency of occurrence of each word in the relevant headline.

The data preprocessing steps we implemented played a crucial role in improving the quality of the dataset and preparing it for further analysis using machine learning algorithms. By removing noise, standardizing text, correcting errors, and removing missing data, the cleaned dataset provided a solid foundation for accurate and reliable news classification.

3.3 Machine learning algorithms

Machine Learning (ML) can be defined as a field that studies the ability of computer systems to learn from data. ML algorithms perform tasks such as creating models on datasets and using these models to make predictions or identify patterns. The main goal of ML is to make decisions or predictions based on future data by generalizing from data. To make these generalizations, ML algorithms try to capture data features and patterns using datasets. In this work, we implemented various ML classifiers to perform the task of news classification based on preprocessed data. The classifiers used in this work include Naive Bayes algorithms and other popular ML algorithms.

3.3.1 Naive Bayes algorithms

Naive Bayes is a probabilistic classification algorithm that applies the Bayes theorem with the assumption of feature independence given the class. The mathematical process of Naive Bayes can be explained as follows:

3.3.1.1 Bayes theorem

The Bayes theorem is a fundamental concept in probability theory. It states that the posterior probability of a hypothesis (class) given the observed evidence (features) is proportional to the product of the likelihood of the evidence given the hypothesis and the prior probability of the hypothesis:

$$P(C|X) = (P(X|C)^{*}\ P(C))/P(X) \tag{1}$$

Here, P(C|X) represents the posterior probability of class C given the observed evidence X, P(X|C) is the likelihood of observing evidence X given class C, P(C) is the prior probability of class C, and P(X) is the evidence probability.

3.3.1.2 Independence assumption of features

Naive Bayes assumes that features are conditionally independent given the class. This assumption simplifies the calculation of probabilities by evaluating each feature independently.

3.3.1.3 Class prior probability (P(C))

The class prior probability represents the likelihood of each class occurring in the training data. It is calculated by dividing the frequency of each class by the total number of samples.

3.3.1.4 Likelihood (P(X|C))

Likelihood represents the probability of observing the evidence (features) given a specific class. In Naive Bayes, likelihood is calculated by assuming feature independence and applying probability distributions specific to each feature type (e.g., Gaussian, Multinomial, and Bernoulli).

3.3.1.5 Posterior probability (P(C|X))

Posterior probability represents the probability of an example belonging to a specific class given the observed features. It is calculated using the Bayes theorem.

3.3.1.6 Classification decision

The final classification decision is made by selecting the class with the highest posterior probability as the predicted class for a given example.

The mathematical operations specific to each Naive Bayes variant (e.g., Gaussian Naive Bayes, Multinomial Naive Bayes, and Bernoulli Naive Bayes) involve estimating class prior probabilities, calculating probabilities based on specific probability distributions and applying Bayes' theorem.

3.3.1.6.1 Gaussian Naive Bayes

It is a Naive Bayes algorithm used to classify data with continuous features. This algorithm assumes that the features are normally distributed and performs probability calculations based on this distribution. The assumption of independence between features is preserved [14].

3.3.1.6.2 Multinomial Naive Bayes

It is a Naive Bayes algorithm used for classifying data with categorical features such as text classification. This algorithm assumes that the features have a multinomial distribution and performs probability calculations based on word frequencies. The assumption of independence between features is preserved [15].

3.3.1.6.3 Complement Naive Bayes

It is a Naive Bayes algorithm that performs well on imbalanced datasets. This algorithm takes the complement of negative examples to compensate for the imbalance between classes. Complement Naive Bayes performs probability calculations by considering class distributions while maintaining the independence assumption between features [16].

3.3.1.6.4 Bernoulli Naive Bayes

It is a Naive Bayes algorithm used to classify data with binary features. This algorithm assumes that the features have a Bernoulli distribution, which represents the presence or absence of features. The independence assumption is maintained, and the probabilities of the features are calculated [17].

3.3.2 Other popular ML algorithms

3.3.2.1 Logistic regression

It is a classification method used to classify samples in a dataset into two or more classes. Using the characteristics of the samples, the logistic regression model estimates the probabilities of belonging to the classes. These predictions are then evaluated on a decision threshold, and the samples are assigned to classes [18].

3.3.2.2 Random forest

It is an ensemble of many decision trees. Each tree is trained with a randomly sampled dataset and used as a decision tree. The random forest makes a classification decision based on the majority vote of these trees. Thus, more stable and accurate results are obtained [19].

3.3.2.3 Linear support vector machine: Linear SVM

It is a classification algorithm used to classify samples in a dataset into two or more classes. By analyzing the features of the samples belonging to the classes, it creates a hyperplane and performs the classification of the points on this plane. Linear SVM works on the principle of maximum marginal separation and tries to provide the largest margin between classes [20].

3.3.2.4 Multilayer perceptron: MLP

It is a classification model based on artificial neural networks. MLP, which has multiple hidden layers, analyzes data features and makes classification decisions using the correlations between nodes or neurons in each layer. MLP is trained and optimized with a back-propagation algorithm [21].

3.3.2.5 Decision tree: DT

It is a tree structure used to classify instances in a dataset into classes. Based on the characteristics of the dataset, decision nodes, and leaf nodes are created, and decisions are made between these nodes. The decision tree reflects a specific hierarchical structure of features and classes and provides a simple, straightforward classification model [18].

3.3.2.6 K-nearest neighbors: KNN

It is a method that considers the k-nearest neighbors to classify an instance in a dataset. Based on the features of the instances, the classes of the k-nearest neighbors are examined to classify an instance, and a class prediction is made based on the majority vote. KNN is a simple and straightforward classification method but can be computationally expensive on large datasets [22].

4. Results and discussions

4.1 Performance metrics

Performance metrics play a critical role in evaluating the effectiveness of machine learning algorithms and making informed decisions. These metrics are important for tasks, such as evaluating the performance of algorithms, guiding the optimization process, reporting results, identifying errors and biases, establishing reference points, and detecting overfitting. In this work, commonly used metrics that are effective in evaluating and comparing algorithm performances, especially for the classification process, are used. Metrics commonly used in machine learning include accuracy, precision, sensitivity, and F1 score. Accuracy represents the ratio of all predictions to all correct predictions and indicates an overall model performance. Precision refers to the proportion of successful predictions relative to all positive predictions of the model. Sensitivity measures the ratio of all positive samples to correct positive predictions and indicates how many true positive samples the model correctly identifies. The F1 score is the harmonic mean used to strike a balance between precision and sensitivity. All metrics have mathematical formulas given by the following equations:

$$\text{Accuracy} = \frac{\text{TP} + \text{TN}}{\text{TP} + \text{TN} + \text{FP} + \text{FN}} \tag{2}$$

$$\text{Precision} = \frac{\text{TP}}{\text{TP} + \text{FP}} \tag{3}$$

$$\text{Recall} = \frac{\text{TP}}{\text{TP} + \text{FN}} \tag{4}$$

$$\text{F1} - \text{score} = 2\text{x}\frac{\text{Precision x Recall}}{\text{Precision} + \text{Recall}} \tag{5}$$

4.2 Results

According to the results given in **Table 2**, the performance of four different Naive Bayes algorithms, Gaussian Naive Bayes, Multinomial Naive Bayes, Complement Naive Bayes, and Bernoulli Naive Bayes, are compared.

The Complement Naive Bayes algorithm has the highest accuracy of 98.31% and was successful in classifying the dataset in the best way. Multinomial Naive Bayes and Bernoulli Naive Bayes algorithms showed high performance with 98.20% and 96.67% accuracy rates, respectively. The Gaussian Naive Bayes algorithm, on the other hand, achieved a lower success rate compared to the other algorithms with 92.92% accuracy.

All Naive Bayes algorithms showed high success in precision values. This shows that the classification results of the algorithms are accurate and precise. The Complement Naive Bayes algorithm has the highest precision value, with a precision of 98.31%.

All Naive Bayes algorithms showed high success in recall values. This shows the ability of the algorithms to accurately identify true classes in classification. The Multinomial Naive Bayes algorithm has the highest precision value with a recall rate of 98.20%.

The F1 score is a combined measure of precision and sensitivity. All Naive Bayes algorithms have high F1 scores. The Complement Naive Bayes algorithm achieved the highest success with an F1 score of 98.31%.

According to the results given in **Table 2**, the performance of six different ML algorithms, Logistic Regression, Random Forest, Linear SVC, MLP Classifier, Decision Tree, and KNN, are compared.

The MLP Classifier algorithm has the highest accuracy of 98.31% and was successful in classifying the dataset in the best way. Linear SVC and Random Forest algorithms showed high performance compared to the other algorithms, with 97.97% and 97.86% accuracy rates, respectively. Decision Tree and KNN algorithms performed poorly compared to the other algorithms with lower accuracy values.

All ML algorithms showed high success in precision values. This shows that the classification results of the algorithms are accurate and precise. MLP Classifier algorithm has the highest precision value with a 98.32% precision rate.

All ML algorithms showed high success in recall values. This shows the ability of the algorithms to accurately identify the true classes in classification. The MLP Classifier algorithm has the highest precision value with a recall rate of 98.31%.

NB algorithms	Accuracy (%)	Precision (%)	Recall (%)	F1 score (%)
Gaussian	92.92	93.08	92.92	92.93
Multinomial	98.20	98.20	98.20	98.20
Complement	**98.31**	**98.31**	**98.31**	**98.31**
Bernoulli	96.67	97.01	96.74	96.77

Table 2.
Results for Naive Bayes algorithms.

The F1 score is a combined measure of precision and sensitivity values. All ML algorithms have high F1 scores. The MLP Classifier algorithm achieved the highest success with an F1 score of 98.31%.

In conclusion, the Complement Naive Bayes algorithm has the highest accuracy, precision, sensitivity, and F1 score, whereas the MLP Classifier algorithm has the highest accuracy, precision, sensitivity, and F1 score. Other algorithms have also shown successful results, but some of them have lower performance.

When the Complement Naive Bayes and MLP Classifier algorithms are analyzed, the Complement Naive Bayes algorithm achieves high success with an accuracy rate of 98.31%. The MLP Classifier algorithm performed at the same level as Complement Naive Bayes with an accuracy rate of 98.31%.

The Complement Naive Bayes algorithm has a precision rate of 98.31%, meaning that the classification results are accurate and precise. The MLP Classifier algorithm has a slightly higher accuracy than Complement Naive Bayes with 98.32% accuracy.

The Complement Naive Bayes algorithm has a precision of 98.31%, indicating its ability to accurately identify true classes. The MLP Classifier algorithm also has a sensitivity of 98.31%, which is on par with Complement Naive Bayes.

The F1 score is a combined measure of precision and sensitivity. Both Complement Naive Bayes and MLP Classifier algorithms have high F1 scores.

According to the analysis results, Complement Naive Bayes and MLP Classifier algorithms show similar performance. Both algorithms have high accuracy, precision, sensitivity, and F1 score values. When making a choice, they can be preferred depending on the characteristics of the dataset and the requirements (**Figures 4** and **5**).

Comparing the results of **Tables 2** and **3**, the overall comparison between Naive Bayes and ML algorithms shows that Naive Bayes algorithms (Gaussian, Multinomial, Complement, and Bernoulli) have an average accuracy rate of 97.77%. ML algorithms (Logistic Regression, Random Forest, Linear SVC, MLP Classifier, Decision Tree, and KNN) have an average accuracy of 87.91%. Naive Bayes algorithms show higher accuracy values compared to ML algorithms.

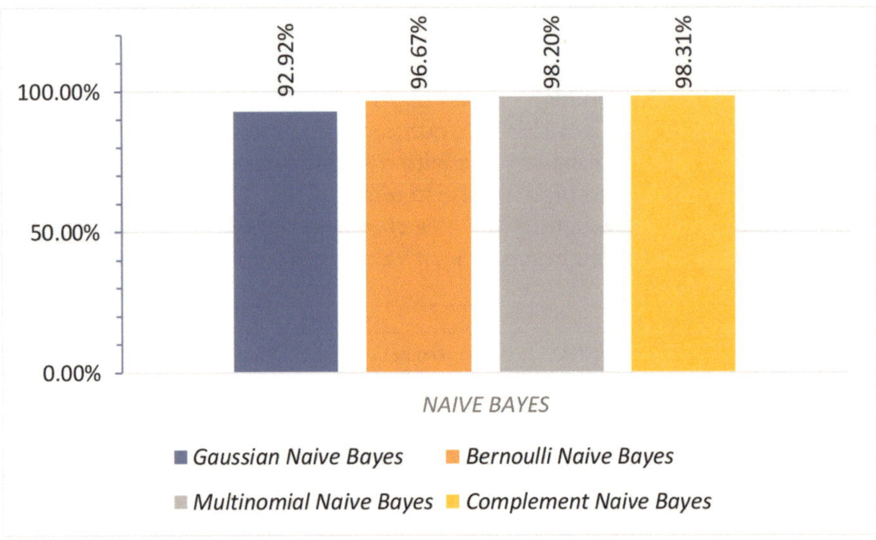

Figure 4.
Accuracy graph for Naive Bayes algorithms.

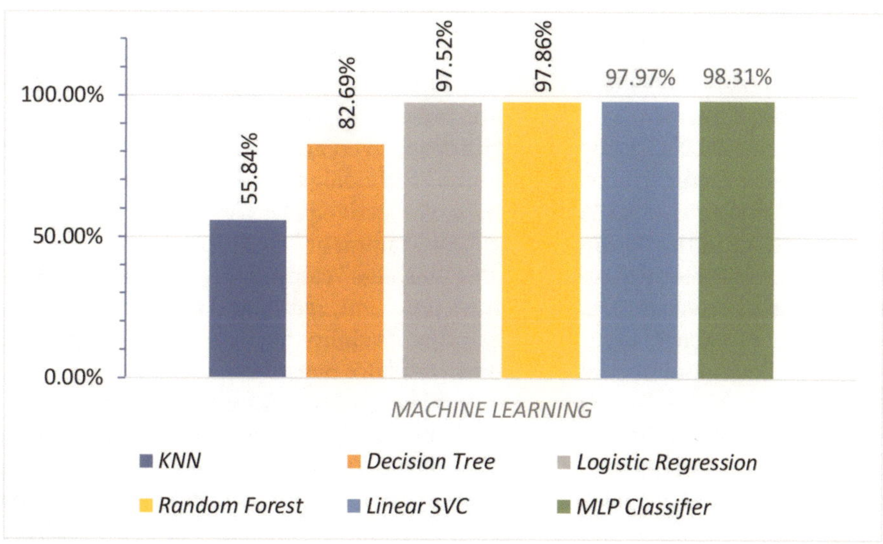

Figure 5.
Accuracy graph for ML algorithms.

ML algorithms	Accuracy	Precision	Recall	F1 score
Logistic Regression	97.52%	97.52%	97.52%	97.52%
Random Forest	97.86%	97.91%	97.86%	97.87%
Linear SVC	97.97%	97.98%	97.97%	97.98%
MLP Classifier	**98.31%**	**98.32%**	**98.31%**	**98.31%**
Decision Tree	82.69%	82.84%	82.69%	82.72%
KNN	55.84%	80.88%	55.84%	56.59%

Table 3.
Results for other ML algorithms.

Naive Bayes algorithms have an average accuracy rate of 97.63%. ML algorithms have an average precision of 84.08%. Naive Bayes algorithms show higher precision values compared to ML algorithms.

Naive Bayes algorithms have an average sensitivity rate of 97.79%. ML algorithms have an average sensitivity rate of 83.29%. Naive Bayes algorithms show higher sensitivity values compared to ML algorithms.

Naive Bayes algorithms have an average F1 score of 97.71%. ML algorithms have an average F1 score of 83.16%. Naive Bayes algorithms have higher F1 scores compared to ML algorithms.

As a result, Naive Bayes algorithms (Gaussian, Multinomial, Complement, Bernoulli) generally outperform ML algorithms (Logistic Regression, Random Forest, Linear SVC, MLP Classifier, Decision Tree, and KNN). Naive Bayes algorithms provide high accuracy, precision, sensitivity, and F1 score values in the news classification problem. However, ML algorithms may also have different advantages in the field of machine learning and may be preferred in certain situations.

The confusion matrix provides valuable information about the algorithms' ability to correctly predict class labels and discriminate between various categories. Based on the Confusion Matrix results of the Naive Bayes algorithms given in **Figure 6**, we can interpret their performance in classifying news headlines into different categories.

The Multinomial Naive Bayes algorithm achieved an overall high accuracy rate of 98.20%. It performed extremely well in the "technology" and "sports" categories, correctly classifying 153 and 194 instances, respectively. However, it made a few misclassifications in the "business" and "entertainment" categories, with only 204 out of 205 instances correctly classified in the "business" category and 157 out of 158 in the "entertainment" category. The algorithm's precision, recall, and F1 score for each category are consistently high, demonstrating its ability to correctly identify relevant examples while minimizing false positives and false negatives.

The Gaussian Naive Bayes algorithm achieved a lower accuracy of 92.92% compared to Multinomial Naive Bayes. It encountered difficulties in correctly classifying some instances, especially in the "work" and "entertainment" categories. For example, it misclassified 11 instances in the "business" category and 14 instances in the "entertainment" category. While it performed well in the "technology" category, correctly classifying 149 out of 161 examples, it struggled in other categories. The algorithm's precision, recall, and F1 score are generally lower than those of Multinomial Naive Bayes, indicating a trade-off between precision and recall.

The Complement Naive Bayes algorithm achieved an accuracy of 98.31%, slightly higher than Multinomial Naive Bayes. It performed consistently across all categories and made only a few misclassifications. It achieved excellent accuracy in the "technology" and "business" categories, correctly classifying all instances. It also performed well in the "sports" category, correctly classifying 196 out of 198 instances. The

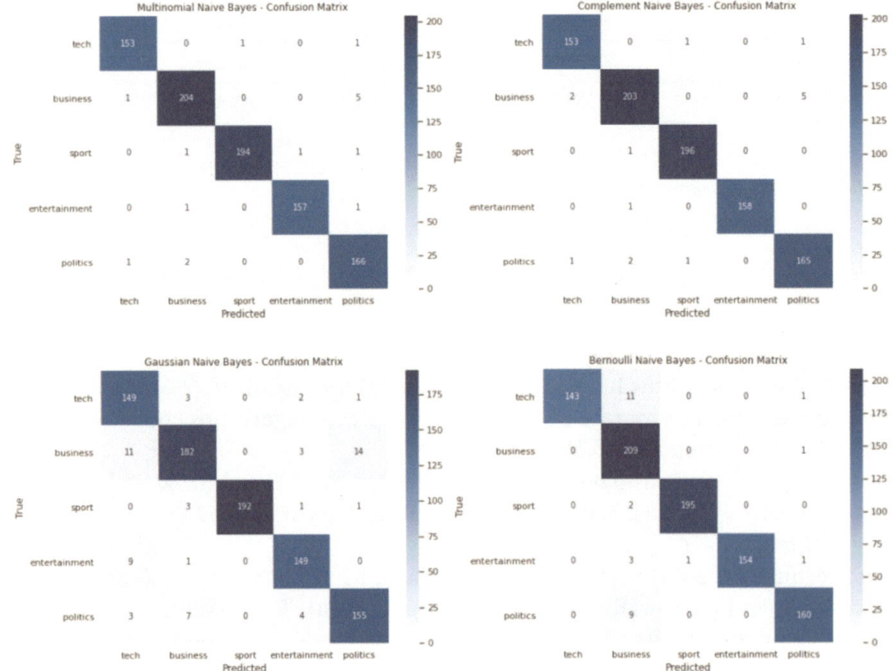

Figure 6.
Confusion matrixes of Naive Bayes algorithms.

algorithm's precision, recall, and F1 score were equally high across categories, making it a robust and reliable choice for news classification tasks.

The Bernoulli Naive Bayes algorithm achieved an accuracy of 96.67%, which is lower than Multinomial and Complement Naive Bayes. It faced challenges in the "business" category, misclassifying 11 instances, and in the "entertainment" category, misclassifying four instances. It performed well in the "technology" category, correctly classifying 143 out of 155 instances, while it faced difficulties in the "sports" category, misclassifying nine instances. The algorithm's precision, recall, and F1 score are generally lower than those of Multinomial and Complementary Naive Bayes, indicating that there is potential room for improvement in its performance.

In summary, Multinomial Naive Bayes and Complement Naive Bayes algorithms performed strongly in classifying news headlines and achieved high accuracy and precision in all categories. The Gaussian Naive Bayes algorithm had some difficulties in discriminating between categories, while the Bernoulli Naive Bayes algorithm performed competitively but faced difficulties in some categories. According to the Confusion Matrix results, Multinomial Naive Bayes and Complement Naive Bayes algorithms seem to be the most suitable choices for accurate and reliable news classification. However, further analysis and finetuning may be required to optimize the performance of each algorithm for specific application requirements.

Based on the Confusion Matrix results of the machine learning classifiers given in **Figure 7**, Logistic Regression achieved a high accuracy rate of 97.52% when we look at their performance in classifying news headlines into different categories. It performed well in most categories, with the highest accuracy in the "technology" category (151 out of 153 correctly classified examples) and the "business" category (206 out of 208 correctly classified examples). However, it encountered difficulties in correctly classifying some instances in the "sports" and "entertainment" categories. The algorithm's precision, recall, and F1 score for each category are generally high, indicating that the algorithm is able to correctly identify relevant examples while minimizing false positives and false negatives.

Random Forest achieved a slightly higher accuracy of 97.86% compared to Logistic Regression. It performed consistently in most categories, correctly classifying a large number of instances in the "technology," "business," and "entertainment" categories. However, it encountered difficulties in the "sports" category, misclassifying five instances. The algorithm's precision, recall, and F1 score were equally high across categories, making it a robust and reliable choice for news classification tasks.

Linear SVC achieved an accuracy of 97.97%, slightly higher than Random Forest. It performed strongly in most categories, achieving the highest accuracy in the "technology" category (151 out of 154 instances were correctly classified). However, it made a few misclassifications in the "business" and "entertainment" categories. The algorithm's precision, recall, and F1 score were consistently high across categories, indicating its effectiveness in news classification.

The MLP Classifier achieved an accuracy of 98.31%, the highest among all classifiers. It performed excellently in most categories, with the highest precision achieved in the "technology" category (153 out of 155 instances were correctly classified) and only a few misclassifications in the "business" and "entertainment" categories. The algorithm's precision, recall, and F1 score were consistently high across categories, making it the optimal choice for accurate and reliable news classification.

The Decision Tree achieved an accuracy of 82.69%, which is lower than other classifiers. It faced difficulties in correctly classifying examples in all categories, with the highest precision achieved in the "technology" category (125 out of 161 examples

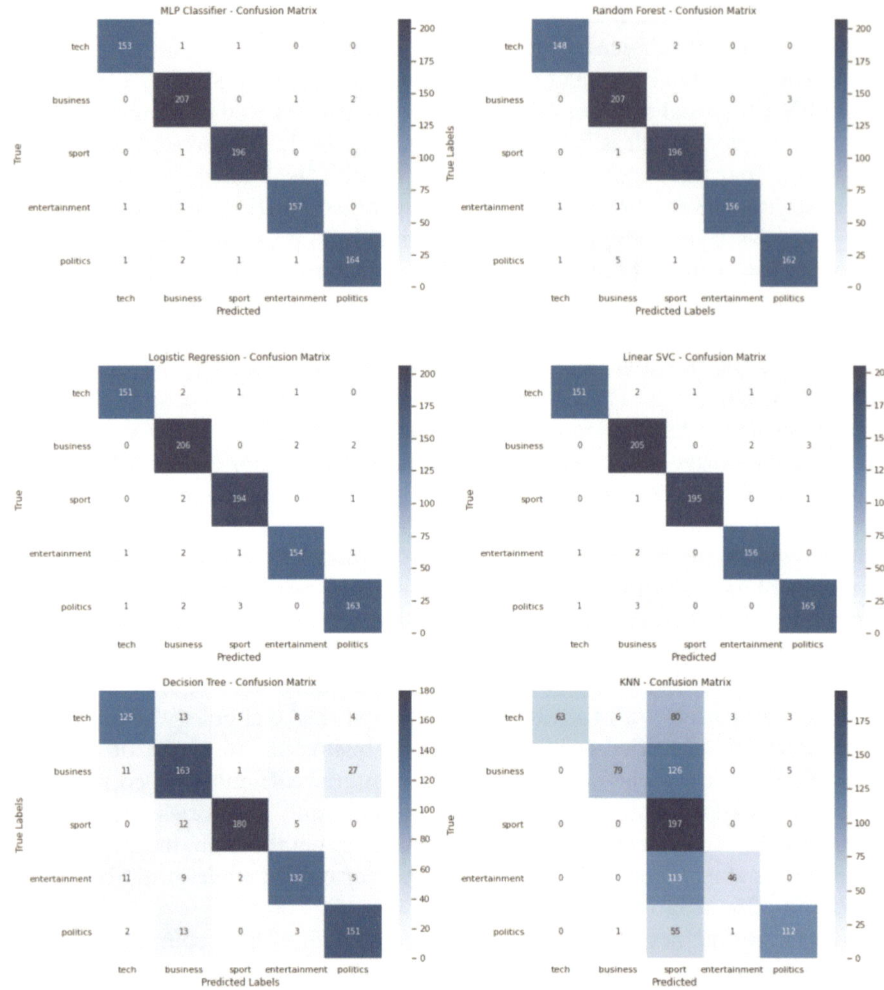

Figure 7.
Confusion matrixes of ML algorithms.

were correctly classified). The algorithm's precision, recall, and F1 score are lower than the other classifiers, suggesting a trade-off between precision and recall.

KNN achieved the lowest accuracy of 55.84% among all classifiers. It faced difficulties in correctly classifying instances in all categories, with the highest precision in the "technology" category (63 out of 152 instances were correctly classified). The algorithm's precision, recall, and F1 score were generally low, suggesting potential room for improvement in its performance.

In summary, the MLP Classifier showed the highest accuracy and precision, making it the most suitable choice for accurate and reliable news classification. Random Forest and Linear SVC also showed strong performance, whereas Decision Tree and KNN had difficulties in discriminating between categories. Classifier selection should take into account the specific context and requirements of the actual research work to achieve optimal and effective news classification results.

In conclusion, the Multinomial Naive Bayes, Complement Naive Bayes, and MLP Classifier algorithms have shown outstanding performance in news classification with accuracy exceeding 98%. Complement Naive Bayes and MLP Classifier emerged as the best-performing algorithms, achieving the highest accuracy, precision, recall, and F1 score. These results highlight the effectiveness of Naive Bayes algorithms and MLP Classifier algorithms in accurately categorizing news headlines, making them viable options for practical applications in content recommendation, information retrieval, and sentiment analysis tasks.

5. Conclusion

In this work, we investigate using Naive Bayes algorithms for news classification using a dataset of news headlines obtained from the BBC News Corpus. The main goal was to build a competent News Classification system and evaluate the performance of various Naive Bayes variants. Through extensive experiments and analysis, we gained valuable insights into the effectiveness of these algorithms in text classification tasks.

Our research findings demonstrate the robustness and effectiveness of Naive Bayes algorithms in news classification. Multinomial Naive Bayes and Complement Naive Bayes emerged as the best performing variants, showing consistently high accuracy, precision, recall, and F1 score. These results are in line with previous research highlighting the effectiveness of Naive Bayes in handling discrete feature data, which is well suited for text-based classification challenges [16, 23].

We also compared Naive Bayes algorithms with other popular machine learning classifiers, including Logistic Regression, Random Forest, Linear SVC, MLP Classifier, Decision Tree, and KNN. MLP Classifier exhibited the highest accuracy among all classifiers. This finding is in line with the recognized ability of deep learning models to effectively capture complex patterns in textual data [24].

Rigorous data preprocessing played a crucial role in the successful implementation of our News Classification system. By cleaning the text data, eliminating stop words, and extracting informative features, we converted the raw headlines into a numerical format suitable for training classification models. Leveraging the Count Vectorizer technique allowed us to skillfully transform text data into numerical representations, thus facilitating the smooth implementation of machine learning algorithms.

Our research adds significant value to the fields of natural language processing and machine learning by providing critical insights into the performance of Naive Bayes algorithms and their comparative analysis with other classification techniques in the news classification domain. The findings emphasize the importance of using appropriate data preprocessing techniques and selecting the most appropriate algorithm to achieve accurate and reliable news classification.

As a limitation, it is crucial to recognize that our work focuses only on a specific dataset of news headlines obtained from the BBC News Corpus. Future research efforts could investigate the generalizability of these findings on larger and more diverse datasets covering various news sources and different topics.

In conclusion, this work proves the effectiveness of Naive Bayes algorithms in news classification and underlines the potential power of the MLP Classifier as an attractive alternative. The insights gained from this research can provide a solid foundation for the development of efficient and scalable news categorization systems with wide-ranging applications in content recommendation, information retrieval, and sentiment analysis.

Author details

Merve Veziroğlu[1*], Erkan Veziroğlu[2] and İhsan Ömür Bucak[1]

1 Iğdır University, Department of Computer Engineering, Iğdır, Türkiye

2 Kütahya Dumlupınar University, Department of Computer Engineering, Kütahya, Türkiye

*Address all correspondence to: mervearg@gmail.com

References

[1] Greene D, Cunningham P. BBC Datasets. 2006. Available from: http://mlg.ucd.ie/datasets/bbc.html

[2] Patel A, Meehan K. Fake news detection on reddit utilising count vectorizer and term frequency-inverse document frequency with logistic regression, multinominal NB and support vector machine. In: 2021 32nd Irish Signals and Systems Conference (ISSC). Athlone, Ireland: IEEE; 2021. pp. 1-6

[3] Saritas MM, Yasar A. Performance analysis of ANN and Naive Bayes classification algorithm for data classification. International Journal of Intelligent Systems and Applications in Engineering. 2019;7(2):88-91

[4] Chen S, Webb GI, Liu L, Ma X. A novel selective naïve Bayes algorithm. Knowledge-Based Systems. 2020;**192**: 105361

[5] Powers DM. Evaluation: From precision, recall and F-measure to ROC, informedness, markedness and correlation. arXiv preprint arXiv: 2010.16061. 2020

[6] Mahesh B. Machine learning algorithms-a review. International Journal of Science and Research (IJSR). 2020;**9**(1):381-386

[7] Rana MI, Khalid S, Akbar MU. News classification based on their headlines: A review. In: 17th IEEE International Multi Topic Conference 2014. Karachi, Pakistan: IEEE; 2014. pp. 211-216

[8] Shahi TB, Pant AK. Nepali news classification using Naive Bayes, support vector machines and neural networks. In: 2018 International Conference on Communication Information and Computing Technology (ICCICT). Mumbai, India: IEEE; 2018. pp. 1-5

[9] Chy AN, Seddiqui MH, Das S. Bangla news classification using naive Bayes classifier. In: 16th International Conference on Computer and Information Technology. Khulna, Bangladesh: IEEE; 2014. pp. 366-371

[10] Bracewell DB, Yan J, Ren F, Kuroiwa S. Category classification and topic discovery of Japanese and English news articles. Electronic Notes in Theoretical Computer Science. 2009;**225**: 51-65

[11] Albahr A, Albahar M. An empirical comparison of fake news detection using different machine learning algorithms. International Journal of Advanced Computer Science and Applications. 2020;**11**(9)

[12] Sristy NB, Somayajulu D. Semi-supervised Learning of Naive Bayes Classifier with feature constraints. In: Proceedings of the First International Workshop on Optimization Techniques for Human Language Technology. 2012. pp. 65-78

[13] Granik M, Mesyura V. Fake news detection using naive Bayes classifier. In: 2017 IEEE first Ukraine Conference on Electrical and Computer Engineering (UKRCON). IEEE; 2017. pp. 900-903

[14] John GH, Langley P. Estimating continuous distributions in Bayesian classifiers. arXiv preprint arXiv: 1302.4964. 2013

[15] McCallum A, Nigam K. A comparison of event models for naive bayes text classification. In: AAAI-98 Workshop on Learning for Text

Categorization. Madison, WI; 1998. pp. 41-48

[16] Rennie JD, Shih L, Teevan J, Karger DR. Tackling the poor assumptions of Naive Bayes text classifiers. In: Proceedings of the 20th International Conference on Machine Learning (ICML-03). 2003. pp. 616-623

[17] Zhang H. The optimality of naive Bayes. Aa. 2004;**1**(2):3

[18] Hastie T, Tibshirani R, Friedman JH, Friedman JH. The Elements of Statistical Learning: Data Mining, Inference, and Prediction. 2nd edn. New York, NY: Springer; 2009

[19] Breiman L. Random forests. Machine Learning. 2001;**45**:5-32

[20] Cortes C, Vapnik V. Support-vector networks. Machine Learning. 1995;**20**: 273-297

[21] Bishop CM, Nasrabadi NM. Pattern Recognition and Machine Learning. Springer; 2006

[22] Cover T, Hart P. Nearest neighbor pattern classification. IEEE Transactions on Information Theory. 1967;**13**(1):21-27

[23] Yang Y, Pedersen JO. A comparative study on feature selection in text categorization. In: ICML. Nashville, TN, USA; 1997. p. 35

[24] LeCun Y, Bengio Y, Hinton G. Deep learning. Nature. 2015;**521**(7553): 436-444